Approaches to Road Safety

Road transport is on a constant course to maintain and improve driving safety in what is a complex process involving multiple forms of risk. It is time for embracing system technologies that give more meaning to the philosophical phrase "safe system" and *Approaches to Road Safety: Evolution, Challenges, and Emerging Technologies* makes the case for the adoption of safer systems in road transport industries through the embracing of technological paradigms, system attributes and more clearly articulated values.

This book offers an account of the professional road safety enterprise in action in the United States, Australia and elsewhere, covering the contributions of Haddon in the 1960s, the creation of NHTSA, the development of the science of human factors, the enduring philosophy of countermeasures to motor vehicle crashes, motor vehicle safety regulations, occupant restraint and protection, collision avoidance, road and traffic design, driver assistance systems, pre-crash scenarios and vulnerable road users. It addresses the harm caused by roadway collisions, including the strategies of Vision Zero and the safe system. It places the enterprise of road safety within the greater system of road vehicle transportation. It also covers ethical preoccupations besides safety, including the rise of sustainability and its operational challenges, including uneven professionalism. Automation and the rise of driverless vehicles are discussed with these being described as a useful long-term change-agent in transportation systems. The important paradigm of connected vehicles and infrastructure is also described, along with the under-utilized science of human factors. The reader is exposed to a nuanced look at the world of road transport safety through its beginnings to the modern age of automation, allowing them to have a contextualized view of the subject area.

Offering valuable insights, this book will appeal to professionals in the fields of safety, human factors, the automotive industry, traffic control, vehicle standards and regulations, transportation systems and road safety policy.

Peter Sweatman is Enterprise Professor in the School of Engineering at the University of Melbourne, Australia. He has over 50 years' experience in transportation research and innovation, and the application of R&D. That experience encompasses vehicles, drivers and infrastructure and impinges on technology, policy and strategic planning. Peter is the former director of the University of Michigan Transportation Research Institute (UMTRI) (2004–2015), the founder and former director of Mcity (2013–2016) and former Board Chair of ITS America. He is the co-founder of CAVITA LLC, providing research services and policy advice pertinent to the deployment of automated vehicles (AVs), connected vehicles and the driving system functions of

V2X. In Australia, Sweatman founded Roaduser Systems Pty Ltd, a successful freight vehicle technology business implementing heavy vehicle research for the benefit of both the private and public sectors, including resource production, manufacturing industry, road freight transport, vehicle manufacture, safety regulation, highway infrastructure and maintenance, and road transport policy development.

Approaches to Road Safety

Evolution, Challenges, and Emerging Technologies

Peter Sweatman

CRC Press
Taylor & Francis Group
Boca Raton London New York

CRC Press is an imprint of the
Taylor & Francis Group, an **informa** business

Designed cover image: image credited to AYO Productions [ShutterStock ID: 2201491467]

First edition published 2025
by CRC Press
2385 NW Executive Center Drive, Suite 320, Boca Raton FL 33431

and by CRC Press
4 Park Square, Milton Park, Abingdon, Oxon, OX14 4RN

CRC Press is an imprint of Taylor & Francis Group, LLC

ISBN: 9781032769080 (hbk)
ISBN: 9781032775852 (pbk)
ISBN: 9781003483861 (ebk)

DOI: 10.1201/9781003483861

Typeset in Times
by Newgen Publishing UK

Contents

1 Introduction

Road safety is an almost universally recognized concept, of great importance to people all over the world. But it is much harder to pin down from a scientific point of view, and subject to interpretation when it comes to public policy and government intervention. Over the many decades of the road safety era, safety has become second nature to much of the world's population. Anywhere there is driving, for necessity or recreation, it is perhaps thoughtless not to admonish those travelers to "drive safely". However, beneath that surface of common sentiment, lies a tangle of designs, motives, dreams and hard realities. This peculiar mix of engineering prowess – both stationary and moving – is fashioned into a humanistic cavalcade of vehicles; unfortunately it is subject to occasional lapses. Collisions and injuries may result, and a portion of these injuries are fatal. From the point of view of an individual driver, these harmful events are rare. However, a highway patrol officer or a hospital emergency room physician may beg to differ: they may well detect signs of an epidemic.

What stands between road users and risk of injury is the road safety enterprise in all its many facets. Safety has developed over a significant period of time, especially since the mid-20th century. As required, it has accumulated elements of engineering, medicine, and law and calls upon professionals working in the distinct sectors of industry, government and academia. Applied research has played a large role, given the low initial knowledge base and the need for well-informed, timely responses to the harm observed on roadways. All the while, politics, regulation, consumerism and activism have all contributed in giving safety a consistent presence in our society.

In the early days, science and philosophy were more peripheral, given the highly vocational nature of transportation itself and its strong identification with engineering thinking and execution. However, it became apparent that new knowledge concerning human propensities and capabilities was needed; the science of human factors applied to driving was developed and made possible by new forms of driving data. The emphasis was still on the orderly development of countermeasures to known types of crashes.

In the new century, digital technology in vehicles, and increasingly in personal devices, offered real-time "situation awareness" data that could be used to alert drivers of impending collision risks. Such technology is well-suited to assisting drivers of vehicles moving in proximity to other moving vehicles, once the limits of

DOI: 10.1201/9781003483861-1

traditional engineering have been reached. Safety is rapidly assimilating 21st-century technology and the new era of road safety is dependent on the wide deployment of digital technologies, along with the more traditional means. Instead of deploying pre-conceived countermeasures to pre-selected crash types, the human driver may be assisted in lessening potential conflicts well before one of them develops into a collision. And the response may be automatic, reducing our reliance upon human drivers' always being alert and capable.

1.1   THE RATIONALE OF ROAD SAFETY

As the title of this book suggests, road safety has been in a state of constant evolution, right up to the present day. This state commenced in earnest in the mid-20th century, even though basic safety measures were demanded by early-century motorized vehicles. The continuous development of road safety defines the modern era of road safety examined in this book. The road safety era started from a low base in terms of knowledge and credibility, but a sense of urgency was created by the proliferation of horrific injuries on the roads. And the seeming randomness of the toll created fear. This ran counter to post-war optimism, economic expansion and the creation of vast new suburbs that relied on the ownership and use of motor cars. The highly practical starting point was the nature of the damage and the timely search for grassroots measures to prevent it. Thus, the iconic road safety term "countermeasure" was born and became a dominant rationale for the entire enterprise. So too, the close study of crashes became its main source of wisdom.

There was no time to develop the best, or a single, approach to road safety, and countermeasures coalesced around the distinct notions of cars, roads, drivers and the environment. All of these elements of the driving system were found to present opportunities for reducing the frequency of collisions. To further complicate matters, safety found the need to understand not only the crash itself, but also what may have led to the crash, and what happens after the crash. Safety learned to treat crashes as complex, high-energy events occurring in very dynamic settings, and having convoluted backstories.

Much of the evolutionary nature of safety arises from the need for conscientious effort to pull all of these strands together, in order to uncover enough knowledge of causality. Many of the safety fraternity have objected to the idea of crash causality, and some still do. But the safety enterprise remains committed to systematically suppressing collisions and injuries without creating new forms of detriment to society or the economy. Such a sustained effort, complementary across many disciplines and social boundaries, requires evolving sets of knowledge, expertise, and methodologies.

Road safety has developed a wide vocabulary that straddles the underlying dichotomies of safe-to-unsafe, cautious-to-reckless and roadworthy-to-unroadworthy and many others. Most of the tools of road safety have specific and well-known meanings, such as regulations, rules, speed limits, occupant restraints and crash barriers. The violent events at the center of road safety have been variously called accidents, crashes, smashes and more. Safety discourse has evolved so that the term crash is generally preferred to the term accident, but this is perhaps not totally settled.

Such subtle shifts in terminology are indicative of philosophical developments that are otherwise not clearly expressed. Does the word "accident" suggest a randomness, or absence of fault, that may impede the further development of road safety? Or restrict funding opportunities, or constrain the drive for solutions? If so, the word "crash" works better.

Road safety has generally learned to favor neutral language and avoid moral judgment. There are several reasons for this. It became necessary to navigate the guardrails of legality and liability. Potential conflicts of interest became apparent; would automakers wish to constantly blame their customers for causing crashes? Would the judiciary use language of condemnation for traffic offenders when they were accustomed to much more egregious behavior exposed in their courtrooms? Not everyone is comfortable with the idea of crash causation; this sits well with the highway patrol, where evidence of fault carries weight, but not among crash investigators, researchers and policy makers. Instead, safety has learned to entertain multiple precipitating factors for a given crash, plus a variety of contributing factors in the same crash event, or sequence of events. Usually there is a critical event – without which the crash would not occur – but this information may prove unproductive in the systematic study of crashes *en masse*.

One of the key notions in road safety is the idea of risk. For example, aggregation of crashes of a particular type, say collisions involving two cars in urban areas, provides some perspective; if the aggregate number is tied back to total distance driven by cars in urban areas, this becomes a certain measure of the risk of such crashes. The distance driven is one way of representing exposure to that type of crash. Safety has a high propensity to identify specific forms of harm, and quantify their frequency of occurrence. Crash risk might be analyzed by vehicle class, driver characteristic or road class. Many different risk ratios may be partitioned by type of roadway, weather, time of day, age of driver, etc. This will often depend on available exposure data: we may well know the age groups of drivers killed in crashes, but not know how much each age group drives or where they tend to drive.

Experienced analysts are aware of reliable data sources and the extent to which different forms of partitioning will be valid. It is certainly possible to normalize the total number of all U.S. passenger car fatalities by the total mileage driven by this class of vehicle. We could then seek to compare this "crash risk" with that of Class 8 heavy trucks. If the latter number was higher, the safety professional may conclude that the heavy truck is "over-represented" or "over-involved" in fatal crashes. This would probably lead to a search for the causes of such differences. This is how road safety analysis tends to proceed. In these cases it is important to be specific about the risk ratio being calculated: the actual meaning of both the numerator and the denominator. For these reasons road safety analytics is recognized as a specialized field. And the wide range of numerators and denominators that can be mixed and matched quickly escalates into a massive selection of metrics that need to be used wisely and interpreted with sufficient contextual knowledge of road safety. While crash risk is a core concept in road safety, it retains a certain flexibility in application. Its use in safety is less formal than it is in industrial engineering, for example.

Suffice it to say that safety is highly evolved from the utilitarian perspective of investigating and countering harm caused by driving. This admirable "first things first" mentality has taken precedence over the formal building of an academic discipline or a safety industry at the global scale. The key concepts and tools of road safety are still a work in progress; they have been developed to be only as good as they need to be. We are dealing with an assembly of knowledgeable methods, products and services that work together to achieve a desired result. But it may be a stretch to call road safety a system, certainly not in an engineering sense. It is more of an applied philosophy, or a paradigm. It is capable of being called upon, usually by the myriad administrative jurisdictions operated by governments, all around the world. In that sense, it is akin to the business franchise system where similar methods and expertise are applied in many, many places, but not necessarily in exactly the same way.

The need for safety to be "all things to all people" not only neutralizes the vocabulary of safety but also tends to place a filter over the enterprise itself and render it difficult to pin down. If road safety operates like a franchise, who is the "franchisor" – the holder of safety's intellectual property? There is no single franchisor. Institutions distributed through several strata of our economy and government make it their business to curate road safety. Each wishes to be ready to implement techniques that they identify as beneficial to their purview and objectives. In the United States, the National Highway Traffic Safety Administration (NHTSA) is definitely a light on the hill; it knows where road safety has come from and provides wide access to what it has produced. It is also active in mapping its future. At the same time, the automotive industry and its suppliers have learned to be expert in safe design and countermeasures.

Safety innovation in the United States requires collaboration between NHTSA and the automotive industry; some of this activity is regulated and some is cooperative. The avenue of collaboration for safety innovation runs through educational and research organizations. There are over 100 transportation research institutes at American universities, many of which focus on road safety. Although focusing on applied research, these institutes are influenced by the academy's instincts for open publication and peer review. The National Academies provide an annual venue for association in the form of the Annual Meeting of the Transportation Research Board (TRB). While road safety is curated locally by those who need it and use it, its constant state of evolution places a spotlight on the more centralized activities of applied, collaborative research.

This model does not necessarily apply in other countries. For example, Australia no longer has an automotive manufacturing industry. Curation of road safety therefore falls mainly to federal and state government agencies. These agencies in turn network with motor vehicle safety regulators in other regions of the world, particularly the programs of the UN Economic Commission for Europe (ECE) and the EU. Australia introduced its first National Road Safety Strategy (NRSS) in 1992, with a ten-year goal for major reduction in fatalities and serious injuries (by 30%). While this was not achieved, further ten-year strategies have followed, up to the present day. The tactics of the current strategy rely on advanced vehicle-based technologies, enforcement of driver behavior and reductions in speed limits. Similar approaches have been adopted in other countries, and so-called *Vision Zero* principles have now been adopted widely.

Vision Zero combines a very long-term goal of zero fatalities (in Australia's case, by 2050) with a "safe system" philosophy that human error should not result in loss of life. According to Australia's NRSS, "system designers and operators, including engineers, planners, lawmakers, enforcement agencies, post trauma crash care workers and others … share responsibility … for designing a safe system". Vision Zero therefore identifies a cadre of professional roles that are needed to initiate a safe system. This goes beyond safety's current role of professional curation of road safety; it also signals a more global scope for road safety. Vision Zero could represent a strategic move beyond the accumulation of safety knowledge pertaining to the transportation system to more directly influencing the design of the transportation system. It signals a desire to advance beyond the limited utilitarian stance of road safety and contribute to creating a transportation system that is less prone to causing harm.

Road safety has evolved to become a knowledge network that becomes operational through many local nodes; all levels of government throughout the world have road safety responsibilities. These operations focus on formulating influence and action that have a high likelihood of rendering crashes less frequent and less harmful. This formulation is probability-based and is not codified in a general sense, although individual aspects may be standardized and even regulated. It has become clear that road safety's operational nodes, especially those associated with government agencies, are seeing the need to design a safer system. While it is not immediately clear how this will be achieved, safety has relied on its organs of innovation to evolve this far, relatively quickly. So we have reason for optimism.

At its best, road safety is a confederated research collaboration working with a vast web of operational sites. The research collaboration may be organized nationally, provincially or regionally and research programs of rigor are the main driver of road safety's evolution. Dedicated and skilled operational agencies are mainly in the public sector; they are knowledgeable and capable repositories of road safety, benefit from the flow of innovation from the core, and feed back based on their experiences. While there is not a single innovation center or core, it is clear that the operational sites vastly outnumber the collaborative knowledge activities.

1.2 COMPREHENDING THE HUMAN HARM FROM ROAD CRASHES

According to the statistics curated and published by NHTSA, and the latest full-year information available, 42,939 people were killed in road crashes and 2,497,657 people were injured in 2021 (1). Such a large amount of harm, a hangover from multiple generations and so widely known, represents a unique problem in public health. But this harm is far from being universally understood. Professional insiders cannot help but view road crashes as an unacceptably large cause of death and injury threatening the entire population; but this is not apparent to the general population, the vast majority of whom have driving licenses. Statistics provided by the National Safety Council (NSC) (2) highlight the "big three" causes of unintentional death, across all age groups. For the year 2021, those causes are poisoning (including drug over-doses) (102,001), motor vehicles (46,980) and falls (44,686). Motor vehicle deaths potentially represent the broadest base of these important public health issues: fortunately only a minority of the population use recreational drugs, and falls are much more deadly for older age groups. Therefore driving is not the most "dangerous" public health threat but could be the most pervasive, even though a small proportion of the driving population tends to have most of the crashes.

The insider's view of high risk in driving is not shared by U.S. drivers, who show no sign of limiting their exposure out on the road: the incredible number of miles driven each year in the United States may be the most stunning aspect of the road safety problem. U.S. drivers covered an astounding 3.14 billion miles in 2021 (1). Because this number is so large, the fatality rate on a per-mile basis looks reasonable. The reported national fatality rate is 1.37 per 100 million Vehicle Miles of Travel (VMT). This is where cognitive dissonance may occur: more than 40,000 lives lost from a single activity, in one year, is a very large toll; on the other hand, an individual could drive many lifetimes before their risk of death would compete with other life concerns.

The number "3.14 billion miles" represents a staggering amount of distance travelled; it equates to 126,103 trips around the Earth in a single year, or 345 circumnavigations per day. That is what turns a very small risk per mile into a national public health issue. Road safety in the United States has focused on steady reductions in that risk factor. Between 1980 and 2000, the rate reduced by 54%; in the following 20 years, the curve of reduction slowed to 12%, but the fatality rate did continue to improve. This could be interpreted as road safety boosting the safety of the road system, but not necessarily that of society as a whole – because people are choosing to drive more. Second only to driving risk officially expressed as fatalities per 100 million VMT is the use of the rate per 100,000 population. While the rate per mile tended to make sense in a narrow view of the engineering system made up of vehicle, driver and roadway, it makes a lot less sense in a city where an increasing amount of harm is visited upon pedestrians and cyclists. The population-based rate is therefore attracting more and more credence. Over time, U.S. road safety has been successful in reducing fatalities per 100,000 population, but not to the same extent as the per-mile rate.

Part of the equation of abundant U.S. driving is the 233 million people with a valid driving license, as of 2021. This is a big factor in the normalization of driving as a public health risk. For example, the country experiences 19.3 annual road deaths per 100,000 licensed drivers. If we narrow the field down to transportation workers who drive in the course of their daily employment, there were 13.3 fatalities per 100,000 transportation workers. And the NSC names transportation as one of the most dangerous occupations, exceeded only by agriculture and mining.

But those who, in aggregate, drive so abundantly for their own reasons are not engaging directly in employment, nor is most of their driving recreational. Driving is rarely an end in itself. To take this into account, economists may describe driving as a derived demand. That is, we choose to drive in support of our daily lifestyles, including access to employment, education, sports, medical treatment, care of others, spiritual needs, essential supplies, recreation, shopping, entertainment, and many others. There is no evidence that many consider daily driving for access to be unsafe, even though the same people would probably baulk at working down in the mines. This is a remarkable contradiction. If daily driving for access was a form of employment, it could well be shunned as being too risky. The terror experienced by many parents when their teens obtain their driving license is well-founded, but it tends to melt away over time because the risk per mile is so low.

Road safety has worked to reduce fatalities – professionally and diligently – for decades, with a good deal of success. The ground transportation system is measurably safer, although professionals know that the current status is not good enough. This does not deter an oblivious populace, free to drive vast distances throughout their lives, motoring endlessly through an imperceptible cloud of risk upon the American highway network. To a lesser extent, this is what occurs in hundreds of other countries. That is why the job of the road safety enterprise is only half done.

1.3 A WORKABLE ENTERPRISE OF ROAD SAFETY

Road safety as we know it is a strongly utilitarian endeavor. That is, it focuses on the consequences of the collisions that may occur as people drive the nation's road network. These consequences include injury and death. Even though such harm is the starting point for road safety, the transportation system is also consumed by the provision of daily access to citizens, the movement of goods and so on. From its functional origins, and right through its evolutionary process, the enterprise of road safety has tried to fix known problems and has steered away from the theoretical and the visionary. So far, this modest, unassuming philosophical bent may well be its most revealing characteristic, along with the accumulated gravitas of professionalism.

Multiple disciplines have come together, over a period of many decades of shared knowledge and common utilitarian focus. This has occurred in a remarkably similar manner all over the world. Such a motivated form of professionalism exerts a noticeable influence over systems of transportation engineering that bear such uncanny resemblance in every country, province and city. The most obvious of these engineering systems are composed of roads and vehicles. And the functional stance of road safety is demonstrated in its mantra of the proven countermeasure. The influence of these countermeasures tends to fall close to the tree. That is, the observed failures of vehicles influence next-generation vehicle design; and roadway defects influence future road design, over a much longer time frame. The errors and transgressions of drivers are more difficult to prove in retrospect, but they influence restrictions placed on driver behavior. In other words, countermeasures stay in their lane – within the specialized contexts of the vehicle, the roadway and the driver.

The valuable progress made by studying crashes and working backward to find solutions should never be underestimated. Each small crash scenario, understood and countered, represents one more pixel in our picture of road safety. But that picture is static, a snapshot of each crash type. It is not possible to wind back the videotape and see the precipitating actions and events. Crashes are system failures. If we can learn more about those failures, we have the chance to create a safer system.

Despite such limitations, the scale of the practical enterprise of road safety is seen today in its influence upon over 1 billion vehicles, owned and driven by over 1 billion people, and millions of miles (or km) of engineered roadways, in over 200 countries. Not only does safety impact engineering practice, at both the strategic and product levels, but it has a pervasive effect on roadway operations, driver behavior and the emergence of system intelligence. Road safety policy, based on carefully accumulated research, evidentiary data and evaluation methodologies, drives these developments. It energizes global spheres of transportation engineering, well-informed policies, and technological systems. It is an *advanced form of philosophically based professionalism* and it embodies the noble aim of eliminating human harm from driving.

The shared proficiency of road safety has borrowed from the many halls of medicine, engineering, technology, law, economics, social science, education and public policy. These disciplines, and others, engage in the design, construction, operation, governance and maintenance of the transportation system. A very great number of these fraternities have some role in road safety, although its individual practitioners are not so engaged full-time. In addition to their expertise, they bring the ethics and mores of their home profession. Those that are employed full-time in road safety tend to reside in federal, state and city transportation departments, research institutes, standards organizations, state and local law enforcement, and private sector companies such as automakers, Intelligent Transportation Systems (ITS), automotive suppliers, management consultants, safety equipment, and training. Safety has managed to develop a cohesive presence through artful association. Critical roles are played by a large number of conferences and standards efforts where voluntary information exchange and debate take place. These activities bring about a certain normalization of views and updating of state-of-the-art knowledge.

NHTSA has played a central role in the professionalization of road safety in the United States. It is dedicated to road safety, operates critical research programs in cooperation with leading research institutes, and has the power to regulate motor vehicle design and driver behavior. It works with research organizations like Virginia Tech Transportation Institute (VTTI) and the University of Michigan Transportation Research Institute (UMTRI). Important activities for safety professionals are the conferences and standard committees of the Society of Automotive Engineers Inc. and the Institute of Transportation Engineers. The Human Factors and Ergonomics Society has played an important role in developing the science of human factors for application in road safety, among other important topics. Researchers in large numbers are attracted to the TRB's Annual Meeting which fosters road safety research, among other topics. There are many well-credentialled safety conferences such as the Driving Assessment Conference. The Stapp Car Crash Conference, in its 68th year in 2024, is rigorous in its exposition of knowledge and techniques in injury biomechanics applied to ground vehicle collisions.

In Europe, large-scale professional road safety programs are curated by the European Union and the UN ECE. The UN ECE operates the World Forum for Harmonisation of Vehicle Regulations (Working Party 29) with the expressed goal of continuously improving vehicle safety while permitting innovation and facilitating global trade. National research institutes such as the French Institute of Science and Technology for Transport, Development and Networks (IFSTTAR), the Dutch SWOV, the German BASt and the U.K. Transport Research Laboratory (TRL) collaborate on road safety research, among other topics. Numerous strong institutes are affiliated through the Forum of European Road Safety Research Institutes (FERSI) and the Forum of European National Highway Research Laboratories (FEHRL). Important conferences include the Transport Research Arena (TRA). Vehicle safety standards operate effectively at the national level; for example, in Germany the DIN standards are applied and organizations like TUV and DECRA provide rigorous testing services.

This high level of European professionalism carries through to important international organizations based in Europe, including the Organisation for Economic Cooperation and Development (OECD) and its International Transport Forum (ITF). The ITF operates IRTAD, a reputable source of road safety statistics covering all developed countries. The International Standards Organisation (ISO) operates out of Geneva and the International Roads Assessment Program (iRAP) is based in London.

Professional curation of road safety has therefore developed noticeably in the fields of vehicle safety standards and safety statistics. Over a long period, convergence of standards for vehicle crashworthiness and occupant protection has occurred. This process is well-supported by national and international bureaucracies and is called international harmonization of vehicle rules. As the newer vehicle standards have migrated to crash avoidance rather than the provision of occupant protection when crashes inevitably occur, the harmonization process has been less clear-cut. Rules for crash warning technologies are more difficult to introduce in a timely manner and take greater account of driver characteristics which may be considered to vary between jurisdictions. Organizations like the OECD have experience in publishing a wide variety of national crash statistics that may reveal trends in the short term and in the long term and tend to facilitate national comparisons. There are strong differences between countries, and the United States tends to be at the poorer end of the OECD's tables.

It is much more difficult to assess and standardize the safety of roads. However, certain considerations in roadway classification, geometry, intersection treatments and signage have been shared internationally. This is a much slower process than vehicle harmonization, with lower expectations of positive impact. While road and traffic rules tend to be similar across borders in a qualitative sense, driver behavior may be expected to show cultural differences. Nevertheless, the fact that everyone is driving highly standardized vehicles on somewhat familiar roadways, and under predictable types of rules, may cause certain convergence in driver behavior.

Despite the impressive arsenal of professionalism that has been directed at road safety, and extensive international attention, the humble utilitarian ethic of safety still dominates. Statistics of crash and injury lie at the center of safety knowledge. Sometimes exceedingly complex, and requiring great skill to extract, manipulate and apply, these metrics are entirely about the harm that occurs while driving takes place. The metrics also try to quantify exposure out on the road, but this type of data is much more limited in scope.

Road safety professionals, and the public officials that employ them, are motivated by trends in the statistics. Are fatality rates this year lower than last year, or ten years ago? Are the rates observed for some classes of road user changing faster than the rates of others? This type of scrutiny has been formalized around the world with the establishment of road safety observatories. The European Road Safety Observatory (ERSO) is a good example. It is managed by the EU and supported by several national road safety institutes, including the National Technical University of Athens and SWOV in the Netherlands. ERSO sets out to measure progress toward the reduction of road deaths and injuries in Europe, identify safety problems and evaluate road safety measures – in addition to exchanges between countries. This is important in Europe because much safety expertise resides at the level of individual nations.

We are now seeing increasing impatience with the unassuming and reactive nature of road safety. Many countries now have national road safety strategies which are developed by leading safety officials and endorsed at the highest political level. Rather than looking for weak spots in crash statistics, these strategies are ambitious in reducing total harm. They adopt targets that call for fractional, rather than marginal, decreases in fatalities and injuries. Such targets have set timelines which tend to be decadal rather than annual. This approach shows a concern for the whole system of ground transportation, and a recognition that system changes take time. The specific tactics recommended in support of these safety strategies are strongly oriented to risk reduction, rather than exposure reduction. That is, they are more concerned with improving the system than with reducing its level of use. Examples of these tactics include speed reductions, crash avoidance systems in vehicles and better driver behavior.

So far, these ambitions are not being met and progress remains incremental, and in some respects stalled. But the focus on "the system" is much needed, and it is extremely challenging. This book therefore devotes considerable coverage to the emerging technological approaches that do speak to the system behavior and quality. These include ITS, connected vehicles (CVs) and infrastructure (Vehicle-to-Everything [V2X], or Cooperative ITS [C-ITS]), and vehicle automation.

1.4   COMPLEXITY OF THE TRANSPORTATION SYSTEM AT LARGE

Any significant changes in road safety need to navigate the realities of road transportation itself. Plus, the practice of safety will need to accommodate foreseeable changes in transportation. Transportation has many purposes and is conducted in a myriad of ways, using engineering systems, cyber systems and others types as required. Some ways are entrenched, and some are opportunistic. The different modes of transportation – traditionally relying on road, rail, water or air – have long been used to help classify and describe the practice of transportation. Transportation infrastructure, including roads, bridges, tunnels, ports, waterways and airports, also provides a definitional impression: the infrastructure provides a lasting physical imprint of the way transportation functions. Because it operates in the public domain, transportation infrastructure is subject to extensive government controls and oversight.

The ability to move people and goods is one of the foundations of human civilization and transportation has adapted and evolved through the ages. From the perspective of road safety, we will not dwell on the economic and social benefits of transportation – the up-side of the transportation "system of systems". Rather, transportation is often studied in terms of its negative, or unintended consequences. In the modern world, these are subject to many interventions and controls exercised by governments. In examining road safety, we will concern ourselves with ground transportation that takes place on the public infrastructure of highways, roads and streets and subject to pervasive informational and traffic systems. Such transportation uses many types of vehicles and personal devices, including cars, light trucks, heavy trucks, buses, motorcycles, bicycles and scooters (the latter may be powered or unpowered). And pedestrians are a critical component that uses no vehicles at all.

Harm to pedestrians, who are clearly not vehicles or machines, is of major importance in road safety; they are considered to be covered by the above description of ground transportation because they are harmed by the actions of vehicle classes listed. Road safety concerns itself with the ground transportation system thus broadly described. Note that transferring people right out of this system to a system of public transport is sometimes regarded as a safety improvement but is not part of road safety. If any part of a person's multi-modal trip uses ground transportation as described above, it is within the purview of road safety.

Daily personal trips in passenger cars, SUVs and light trucks represent a very large part of the annual accumulation of miles driven. As we have said, this high level of driving activity – one of the most prominent measures of exposure in the ground transportation system – combines with the system's crash risk to produce collisions and injuries. This will continue to be the case, and there is no indication that general mileage paring will be seriously considered as a safety measure. What is changing is the interaction between those vehicles and vulnerable road users (VRUs). Urban populations are increasing faster than populations in general and they are more likely to use alternative forms of mobility, relative to cars. VRUs do not fit the traditional patterns of road safety. Less is known about the risks of VRU injury, and very little about their exposure. The old utilitarian approach is not working for VRUs. Pedestrians are not just walking; they are part of a new paradigm of mobility free choice in cities. Part of the motivation for new mobility is widespread opposition to transportation's reliance on fossil energy. This trumps the usual reticence of road safety and may mean that car-free movement is partly forgiven for being less safe. Or that society has moved on and simply expects new modes to be safe. And we have learned that transportation users tend not to be aware of the actual risks attached to their movement on the public roads.

If road safety set out to modify the transportation system in the interests of safety – beyond current practices of modest interventions – what changes may be sought? Given that safety is currently mainly concerned with risk reduction, would changes in exposure be placed on the agenda? Beyond the ebbs and flows of the demand for driving – brought about by economic factors such as unemployment, fuel price and so on – what wholesale measures would be considered to alter where people choose to drive and move, and how much they drive (and move)? Pricing is used in some mega-cities around the world to control the number of cars and other vehicles trying to enter the center. Such schemes are aimed squarely at reducing traffic congestion and gridlock. Lane access schemes are used on busy freeways in certain parts of the country, either through vehicle occupancy requirements, usage-based payment or vehicle eligibility requirements (for example, electric vehicles).

But it is unlikely that road safety and its political masters would be attracted to these sorts of measures, especially any scheme that offers greater safety to those who pay more. The most likely candidate for system-changing safety would be vehicle automation – so-called driverless cars. In fact much of the vigorous public sector support for driverless cars is predicated on the idea of bypassing human error as a major cause of crashes. If deployed on a sufficient scale to transform road safety, driverless cars would represent a major existential change in the system of ground transportation. On the other hand, partial use of technologies used in driverless cars is already occurring and represents a continuation of current vehicle-based systems for crash avoidance. As such, they are crash countermeasures within the current transportation system – not yet system-wide changes.

Deployment of driverless vehicles will usher in an important transformation of the transportation system, affecting the entire system as we know it. Not only do we need to be concerned about vehicle automation, and especially the automated driving system (ADS) that replaces the driver, we also need to take care of the Operational Design Domain (ODD), infrastructure innovations and human behavior pertaining to occupants of the automated vehicle (AV) and the drivers of other vehicles in the traffic stream. The driverless-enabled transportation system will be expected to be a lot safer than the current system. But our utilitarian approach to road safety may no longer be sufficient to manage such a deep-seated change.

Another important, and related, change to the transportation system is the connection of vehicles and infrastructure. This paradigm is generally known as CVs, V2X or C-ITS. The technologies of vehicle-to-vehicle (V2V) communication and vehicle-to-infrastructure (V2I) communication are designed to detect risks of collision as they build – in real time – and immediately provide warning and intervention. From the perspective of road safety, this technology cuts right to the heart of crash risk as it happens. But it is most dependent on the network effect of all, or many, vehicles in the traffic stream always being "on", no matter where you decide to drive. And completely compatible, foolproof technology platforms are needed in several hundred million vehicles – old and new – in public infrastructure – physical and cyber – and in wireless connection between the two platforms. This is asking a lot, and it is only the beginning: all vehicle classes need to be included and VRUs need to be spirited into the system as well. But the benefits do go beyond safety to benefit energy use, emissions and the environment. V2X is conducive to new traffic management approaches that deal with platoons and streams of vehicles, rather than individual vehicles, offer much wider scope for so-called green waves of vehicles at intersections, and for minimizing stoppages at traffic lights.

In addition to being complementary technologies – AVs work a lot better with the help of V2V and V2I – automation and connectivity are powerful system creators. We are on the cusp of establishing transportation as a recognizable cyber–physical system. But as long as human unreliability can so easily break the chain of cause and effect, we remain reluctant to embrace the transportation system – permanently altered for the better – that lies just over the horizon.

If safety is intending to initiate such major changes in the transportation system, what other change agents are likely to impact that same system? Safety is not the only uber-influence on the design and operation of the ground transportation system. In addition to safety, other negative consequences usually assessed in ground transportation include use of fossil fuel and attendant emissions of carbon, traffic congestion and air pollution. Criticism is also directed to automotive domination of the urban built environment, urban sprawl and global tensions caused by the thirst for oil.

Traffic congestion has been increasing for years and has entered the public consciousness as a daily toll on people's time and communal tolerance. It has become symbolic of the limitations of modern infrastructure and, sometimes, the impotence of the traffic engineering profession. More recently and precipitately, transportation's domination of fossil fuel consumption has caused a revolt against the notion of open-ended expansion of the transportation system, especially the notion of more and more cars. The global politics of energy and climate are powerful enough to bring significant change to the transportation system. After many years of tacit reliance on the curated vehicle, the hierarchical roadway, the educated driver and the moderated traffic stream, all of these tropes are being called into question.

Most notably, the new mode of climate-friendly, healthy personal mobility in urban settings – including many more trips by foot, cycle and scooter – has helped to create the paradigm of the VRU. These personal "micro-modes" do not mix well with motor vehicles. The proportion of persons killed outside the vehicle, through being struck by the vehicle, has increased to the extent of reaching one third of all road fatalities by the end of the 2010s. This trend has partly negated mainstream efforts to improve road vehicle safety through avoidance of collisions between vehicles and occupant protection. Such changes in the transportation system, which in turn have a profound effect on road safety, tend not to be created by professional curation. Rather, they arise in new and largely unregulated marketplaces that may have very different philosophies. In fact, some markets believe in disruption instead of curation; trial and error instead of product testing; personal freedom instead of regulation; conflict instead of cooperation; and equality of access rather than market forces. Such instincts run counter to the curative long-term philosophy of road safety. It is incumbent upon road safety to fully integrate VRUs in its philosophy and practice.

Other new forms of urban mobility have materialized in the new century, appearing to relish urban dwellers' desire for a more immediate way to move and more innovative use of current automotive capacity. Ride hailing and ride sharing have established a strong following but have generated controversy about which other modes they may displace, their effect on city traffic congestion, and whether they represent an ethical form of employment, as well as their effect on safety. While there is little reason to believe that this type of driving is any more or less safe than any other form of urban driving, such operations have created a need for pick-up and drop-off zones. This variation to city infrastructure remains informal, and concerns have been expressed about riders being struck by passing motor cars as they enter and exit ride hailing vehicles.

The influence of energy and climate policies on road safety is a completely different type of question because these are global matters external to the transportation system. Present-day efforts to conserve fossil energy supplies and to substitute the use of renewable energy affect vehicle design significantly but have limited direct influence on roadway design and operations. Battery-electric vehicles (BEVs) have relatively heavy battery systems spread across the floor level but otherwise minimize weight through design and material selection of the vehicle's structure, body and interior. Such design philosophy affects vehicle performance and handling, but there are no clear indications of safety consequences. Early fears of reduced vehicle crashworthiness appear not to have materialized. One unanticipated area of safety concern is battery fires which attract publicity despite their low rate of occurrence. Other new-energy vehicles, including hybrid and hydrogen drive, have a similarly low safety profile.

Global awareness of atmospheric carbon emissions, especially from motor vehicles with internal combustion (IC) engines and from electricity generation, has propelled high political action. It is generally accepted that engineering solutions are required, and such solutions have materialized, possessing transformational impact. Most debate has centered on the cost of innovations like battery electric vehicles (BEVs) and methods of charging. But the field of transportation policy-making has now expanded to include electricity generation. In this important field, non-fossil methods old and new have received much attention. However, the rate of automaker investment in BEVs and the de-commissioning of ageing traditional power stations indicate that a transformation is underway. Regardless, road safety is largely de-coupled from these changes in vehicle systems. Of more interest here are any lessons for changing aspects of the transportation system that do affect safety.

Innovation in energy production has certainly been encouraged by government investment and by incentives offered to consumers. However, legions of users have created the demand through their own beliefs and virtuous behavior in the market-place. Government regulation has played only a small part. Given that perceptions of virtue play a part in road safety as well as in vehicle energy choices, is there any evidence of competition between these two outsize influences on transportation? Cost is obviously a huge factor perennially attached to vehicle energy, or fuel, but consumers will still expect vehicles to be safe, even if renewable energy is going to cost more. It is very unlikely that they would attempt to engage in any trade-off between their stakes in the family's safety and the health of atmospheres and oceans. In contrast to high public awareness of energy and climate change, road safety is not a hot-button issue. The very lack of a true sense of driving risk among most people with driving licenses and motor vehicles remains a huge challenge for the road safety profession.

Moving beyond the prosaic considerations of dollars and cents, we begin to see a new primacy of the energy influence. At a higher level of social approval, including the ability of the family members to lead successful lives and improve the prospects of future generations, climate health would probably outweigh road safety. Not only that, it will set the stage for new safety challenges. Totally new mobility modes and technologies will be invented. Whatever may materialize, the current generation will expect new modes to be safe. We are seeing this with the development of driverless cars; they are expected to be safer than traditional cars – some would say a lot safer. It is therefore unlikely that the safety of the current transportation system per se will continue to be a leading issue for the next generations. This does not mean that safety has a limited "shelf life". Rather the philosophically based professionalism of road safety will need to be handed down, but in a form adaptable to a new diversity in mobility.

Road safety is a remarkable institution: a boundary-hopping professional philosophy that is applicable worldwide and has curtailed human harm from driving. It has done so in the face of an almost unimaginable volume of driving. People the world over drive so much that ground transportation – utilization of cars and trucks – is the largest consumer of energy. In the United States, transportation uses 27.5 quads of energy; industrial production is relegated to second place with 26.6 quads. All this driving, increasing inexorably over time as populations increase and economies develop, presents road safety with a huge challenge. Even though serious crashes seem to be rare events when considered on a per-mile basis, the aggregate of annual road fatalities rises to represent one of the largest causes of unintentional death in the United States. This is very obvious to the profession of road safety, and to all related industries and government agencies. However, the vast cohorts of people who drive do not see it this way. They are generally conscious of taking care when driving, but they do not go out of their way to avoid taking trips, or to minimize their time on the road.

One aspect of the transportation system that causes a strong and despairing reaction from all drivers is rising levels of traffic congestion. The management of traffic congestion, while differing from the safety enterprise in some respects, also relies upon highly associated professionalism, broadly applied in many places. But here the onus is more on the public sector – the transportation departments of states, counties and cities – and their relationship with the traffic control, road construction and ITS industries. Safety remains a central tenet, but traffic engineers have their own problems. What is good for traffic movement is often good for safety, but not always. Reduction of traffic conflict serves as a universal goal, but efficient traffic flow generally equates to higher speed of the traffic stream.

Somewhat in contrast, safety suffers from higher speeds of individual vehicles, more so than the stream itself; divergent vehicle speeds within a traffic stream constitutes a negative influence on safety. Over time, we will continue to see a greater emphasis on traffic efficiency. This is not detrimental to safety but will require a better understanding of the influence of traffic dynamics on safety. Road safety has grown up with a focus of one vehicle and its occupants and a specific roadway location that becomes a crash site. Such considerations will sometimes encompass more than one vehicle, but no more than a two or three – except for severe weather-related events. The evolution of road safety will need to embrace more complexities of system behavior. It will be necessary to adopt the road traffic environment as a specific element of the system of road safety: traffic as a technical source that contributes to crash causation, along with the vehicle, the road and the driver.

The many standards, rules and laws applied for the sake of road safety, and all of the government safety agencies, tend to be channeled. They target either vehicles, roadways, drivers, VRUs or traffic. And they represent separate streams of expertise and practice. The net effect on the nature of the transportation system is not well articulated, but it emerges over time – another example of our utilitarian world. It should be noted that vehicles and roadways have been dealt with in a professional manner for many decades. Roadways are designed to accommodate vehicles. Vehicles have evolved much faster than roadways and drivers quickly adapt to new levels of vehicle performance. Evidence of adaptation in the system may be seen in highway operations; for example, traffic streams adopt shorter vehicle headways than in the past and maintain them at higher speeds of travel. This type of system behavior is not planned but may be observed after the fact. Generally, vehicle designers would expect road standards to be kept stable over time but may expect great diversity and adaptation among their own customers – the drivers.

Government regulation has a certain reputation for stifling innovation. For highly regulated industries like the automotive industry, compliance with rules and regulations is a foundational activity and requires an appropriate allocation of resources throughout the company. The need for re-certification of new designs is never far from mind when companies weigh customer-pleasing developments against innovation costs. At the system level, regulations have not prevented steady proliferation of new vehicle classes and types: SUVs have almost replaced sedans and people movers have attracted a significant portion of the light vehicle market. Meanwhile, light trucks have increased their appeal and expanded into crew-cabs – not for the crew, but for the family.

Roadways are much slower to evolve, having very long lifecycles relative to vehicles. The post-war expansion of national economies, and the relative freedom of suburban living, caused serious examination of requirements for new roadways. This led to a classification system according to purpose. Urban redevelopment occurred throughout the United Kingdom following the destruction of World War II, bringing the issue of burgeoning vehicle traffic into sharp focus. In the 1960s, British urban planners pioneered the study of urban buildings interacting with land use and transportation and promoted active consideration of "the flows of people, vehicles and goods" that are generated by certain layouts. This led to the notion of designed urban transportation systems. A landmark report by Buchanan (3) sought to manage, and even pre-empt, the congestion problem through certain restrictions on traffic combined with stratified access to infrastructure in cities. This type of thinking, and the use of the term infrastructure, implied a form of system design: that is, the nature of the infrastructure would cause traffic to behave in a more systematic manner. Different classes of roadway were easily recognized from a physical perspective: the visual typologies of freeways, arterials, feeders, boulevards and local streets were readily assimilated by citizens. These forms of infrastructure clearly have had – and continue to have – a lasting effect on our system of ground transportation.

1.5   EVOLUTION OF ROAD SAFETY

Road safety is not concerned with natural selection, nor survival of the fittest. It sets out to negate harm in the road transportation system, and it starts by observing and quantifying motor vehicle harm and its causes. If biological evolution makes us more resistant to vectors that seek to cause harm, safety first wants to deflect those vectors. Then it wants to use those learnings to make the transportation system more resistant. Safety therefore sounds like a multi-phase endeavor that is deliberately self-conscious and constantly seeking to remake itself. It could be said that the evolution of road safety has taken it from countermeasures for injuries to car occupants – whenever those cars are involved in collisions – to the removal of risk in the road transportation system. The former case considers the vectors of injury – in relative isolation from the road transportation system as a whole – whereas the latter case seeks to inoculate the system (human/physical/cyber) against risk. The arc of road safety has therefore taken us from graphic reality to intellectual prescription. Its scope is much, much broader but it now entails levels of abstraction that may obscure safety's mission.

The founding fathers of modern road safety, epitomized by William Haddon (4), saw motor vehicle crash injury as an epidemic. The scientific methodology of epidemiology was therefore required to prevent the interior of the car from injuring the "host" – the driver and other occupants. It was not considered feasible to prevent collisions from occurring in the first place because humans were too unreliable and error-prone. But that attitude changed immensely once the new science of human factors was developed and then bolstered with digital technologies. Road safety was able to evolve beyond the crashworthiness of the vehicle to the invention of new vehicle systems that would avoid crashes.

These hard-won gains are now threatened by increasing numbers of serious injuries and fatalities among urban dwellers who were choosing to move in unpowered modes, including walking. This is a throw-back to the early days of humans being injured by a vector – car interiors. But now the vector is the car exterior and many more people outside cars are being harmed. Part of the reason for people favoring unpowered movement is a growing perception of highly powered vehicles dominating and degrading populated areas. In addition to a new form of harm that is more difficult to counter than previous forms of motor vehicle injury, VRUs perhaps bring evidence that the center of road safety is not holding. People's emotional intelligence is not invested in road safety; it is invested more strongly in climate change. And that interest is not short term, or opportunistic: it is future-framed right across the 21st century.

The utilitarian origins and trajectory of road safety have not inspired popular admiration or faith in its future development. And safety's "in-house" branch of science – human factors – has been underutilized for a very long time and has had surprisingly little influence on the functional design of the transportation system. Major changes in the transportation system are going to be needed on safety grounds. Meanwhile, those changes are not widely appreciated and generational changes in urban movement are causing more problems. We have endured a long gestation period; now a compelling and systematic resurgence of safety is needed. The technologies of AVs are capable of detecting risks of moving vehicle collision and acting to immediately remove the risk. CV and infrastructure technologies can prevent collisions at intersections of roadways and merges of traffic streams. A significant public–private commitment is needed because these technologies only work when the vast majority of vehicles and infrastructure traffic locations are equipped with the technology, and it is always on and always operational. Further, the technology must remove the possibility of any vehicle striking a VRU; at the same time it is reasonable to expect the urban commons to be designed to minimize the exposure of VRUs to vehicle strikes.

Road safety cannot meet such a set of challenges without adopting a new philosophy. Hitherto, road safety was predicated on cancelling specific types of crashes, virtually one by one. Road safety defined itself retrospectively by studying and then countering many types of crashes. But would it be possible to step back from the brink and study conflicts that mostly don't resolve themselves in an actual crash? A new wave of technologies that could connect vehicles wirelessly, plus machine vision, radar and lidar, could even bring about conflict-free driving. Road safety would no longer be defined by traffic conflicts that become crashes but would become an embedded virtue of the road transportation system. These emerging technologies are hallmarks of the next phase of road safety.

Evolution is more than a series of developmental steps. The changes that occur must lead on automatically to further, ever more fundamental changes. The enterprise of road safety exists in its philosophies and knowledge base but only becomes a system when applied to the transportation practice. In that sense, the philosophy of road safety has evolved because utilitarianism could no longer provide the needed rate of change in the transportation system. Cyber, including connectivity and automation, changes the transportation system and creates new ways for it to be safe. As they appeared early in the 21st century, connectivity and automation were welcomed as new and more powerful ways to avoid crashes. In other words they were considered to be countermeasures to known crashes – an extension of safety's utilitarian approach. But if we view the technologies as hitherto-missing pieces of a fully fused road transportation system, that system becomes changeable. Not only for safety and efficiency, but driving could be made more equitable, more secure and more sustainable.

As road safety struggled to maintain its momentum in the new century, revised philosophical principles were adopted by safety leaders and national governments. However, it proved difficult to untangle the new philosophy from the old utilitarian practices. The philosophy of the safe system, where errors would not always have harsh consequences, is widely accepted. But the immediate pivot to numerical targets of fatalities says little about the design of a safe system. Long-standing ethical principles attributed to the great philosophers are overdue for application in road safety. The reader is directed to the Stanford Dictionary of Philosophy (5) for peer-reviewed exposition. The ethical approach of "consequentialism", or utilitarianism, is often attributed to John Stuart Mill (6) in the mid-19th century. His philosophical gold standard was "greatest happiness for the greatest number" (GHGN). In the case of road safety, "greatest happiness" often translates as "least harm" caused in the pursuit of "greatest utility" of driving. There are therefore two sides to the GHGN principle, but road safety has tended to focus on the harm, and the safe system mainly refers to the propensity for harm. The Vision Zero movement goes a little further and says that human harm should not be traded off against economic benefit. As such, it is a quite restricted version of Mill's GHGN. It is much closer to Immanual Kant's philosophy of perfect duty: expecting that all drivers will follow the moral law and never cause deaths on the road.

An essential adjunct to road safety's embrace of 21st-century technology is the development of a broader and more robust set of ethics. In the Old World, philosophers like John Locke had long warned against the brutish "state of nature" and advocated a certain degree of handing-over of administrative powers to a central authority. Locke called this the social contract. In the 20th century the idea of centrally conferred powers exploded in a way that the earlier philosophers, including Mill, would never have foreseen. A combination of engineering professionalism, economic rationalism and regulation took hold and moved from infrastructure construction to motor vehicles, to traffic rules and their enforcement.

In the 1970s, the American John Rawls came up with contractualism as a more rigorous framework for dealing with the philosophical crown jewels of individual equality and freedom within society. It is noticeable that road transportation, and road safety, has treated equality as a core ethic, although it is rarely articulated. All drivers are subject to the same rules and there is a general sense of "we're all in this together". This fits Rawls' thought experiment where the social contract is negotiated by well-informed representatives who are active in society and are ignorant only of one important thing – their own status in society. The matter of trust is surely baked into the transportation system; this surfaces repeatedly in our confidence that an oncoming driver will take care to avoid a collision. Again, this is rarely articulated but affects people's decisions. For example, the limits of trust may quietly influence the selection of a large SUV as the family vehicle; such a vehicle may be assumed to be safer in a collision with another vehicle.

Road safety is a prime modern example of the social contract. Obviously people's acquiescence to government controls carries over from the many walks of public policy wrapped up in democratic governance. But road safety also entails its own notions of trust and responsibility. Every mile, the many decisions and driving actions are not made from first principles but arise from the "safety contract" that has become part of our culture. The safety contract is experiential and has evolved in order that we have enough mental clarity to make the right decisions most of the time. In making those decisions we take care of ourselves as well as considering the well-being of other road users. Dilemmas are rare because a collision is a collision for both parties. Dilemmas, and increased reaction times, would strongly increase the probability of a crash; but dilemmas survived would strengthen the safety contract. None of this is written down.

The ethics of driving safely is clearly a complex matter. Given the sensitive nature of the subject, and the intrusion of personal morality, it is not surprising that it has been examined only as much as required. When it comes to a collision between two light vehicles, and given the guaranteed occupant protection afforded by all modern vehicles, ethical considerations tend to be moot. Collisions between light vehicles and heavy trucks are more likely to cause rancor, and the one-sidedness of the harm may provoke social or legal consequences for the carrier or the truck driver; for example, the driver may lose his job. A VRU being struck and killed by a light vehicle comes very close to justifying a moral stance – the driver viewed as breaking the moral law.

But even then there are often extenuating circumstances, including pedestrians failing to use the appointed crossings. Every type of road death or injury is susceptible to a sense of "it could have happened to anyone", and a reluctance to condemn or punish.

We have reached the point where deeper examination of driving ethics is needed. This has been brought on by serious experimentation with AVs. While most experts involved in road safety see AVs as being, at least potentially, a lot safer the devil is in the details. How could a team of programmers employed by the manufacturer of the automated driving system (ADS) deal properly with the split-second decision process faced by the AV when a collision is imminent? Matters of concern even extend to algorithms that may choose what type of crash would be the least harmful, and who to collide with first if collision is unavoidable. Of course we know very little about the pre-crash thought processes of human drivers, and the extent to which drivers may quietly try to protect themselves at the expense of others.

Philosophers have invented cartoonish "trolley problems" to illustrate ethical dilemmas in taking action to save certain people and sacrifice others, and to show how different schools of ethics would provide different answers. But one of the key issues in mixing AVs and traditional vehicles driven by ordinary people is that ordinary people cheat and breach rules. If AVs always follow the rules, we no longer have a system totally dominated by utilitarianism; rather, we introduce a strain of deontology. As Chris Gerdes of Stanford and others have suggested, AVs should be programmed to always drive according to the road rules and to maintain a certain margin of safety relative to other vehicles (7). If a person-driven vehicle fails to follow such protocols, the AV will do its best to minimize the collision with that vehicle. However, it will never take evasive action that harms other "innocent" vehicles that happen to be present.

In addition to invoking Kant's categorical imperative – in this case, "always follow the road rules" – road safety in the post-AV world of driving may also introduce a machine version of Aristotle's virtue ethics. Gerdes uses the term "duty of care" to cover the responsibility of the machine, or the person, to drive safely. And that duty of care extends to other vehicles (and of course their human occupants) – and not making a complex decision to involve them in the crash. Drivers that do the right thing consistently – whether humans behind the wheel or driving machines – are "exercising the virtues", as described by Aristotle. Over the modern era of road safety, we learned that humans were not able to do this consistently enough and therefore needed to be coerced and provided with countermeasures. Going forward, driving machines already show the potential for driving safely with greater consistency than humans, and always exercising duty of care. Driving machines are also much more able to adopt straightforward maxims akin to robotics: for example, stop at red lights, stay in the lane and never harm a human being.

Beyond these matters that speak to the ability of a driving machine to consistently out-do a human driver, it is not clear how machine substitution would take place. As a pure safety play, we need to imagine AVs taking over the lion's share of the 3 trillion-miles-plus of human driving. At best this would need to occur in stages but the nation's licensed drivers, numbering in excess of 200 million, are offering little encouragement for stand-alone AVs on a large scale. The safety strategy of

AV substitution would also underestimate the agency of driving in many people's daily lives.

1.6 PROFESSIONALISM IN A COMPLEX SYSTEM

In addition to its reliance on socio-engineering systems, transportation is strongly aided and abetted by national and international cadres of professionals. The numerous disciplines mentioned above all have professional mores and ways of associating. The American talent for association was noted by the pioneering political scientist Alexis de Tocqueville (1841), and he recognized association as a key element in democratic society (8). Learned societies have played a critical role in advancing necessary disciplines, bringing collective expertise to large problems and connecting with otherwise-separate disciplines. It would be easy to underestimate the vital contribution of professional association to the system-of-systems we now know as transportation.

Engineering professionalism came of age in the 20th century, and the field of transportation became an early exemplar. In the first half of the century, the design and construction of roads and highways provided some of the first major avenues of transportation professionalism, showing the way to the collective development of engineering design standards. The establishment of the Bureau of Public Roads, paving the way for the U.S. Department of Transportation, and the Transportation Research Board of the National Academies, cemented the place of the engineering profession at large in transportation.

Tocqueville considered the talent for association, and acts of free association, to be one of the hallmarks of the New World. Societies juggled the powerful ideals of freedom and equality of their citizens. From the creation of the Union, Americans were free to associate and develop the power of collective intent. This was a totally new point of departure, unencumbered by Old World worries about the state of nature and acquiescence to a certain degree of handing-over of administrative powers to a central authority.

Mill (1859) believed that government regulation of man's affairs was only justified if "mayhem was certain", and not if it was merely a possibility (6). Fast forward to the present day, extensive regulation and enforcement of driving is widely accepted – grudgingly wherein some cases. This general acceptance is at least partly assisted by regulations' being generally based on evidence of effectiveness, and demonstration of fair burden of cost. Quite iconically, this requires careful scrutiny of prevailing unsafe consequences versus future consequences as mitigated with the predicted benefit of countermeasures. And once such improvements are in place, the results should be assessed with before-and-after studies.

Safety has had the benefit of professional curation over many decades, sustained multi-disciplinary research and excellent public policy development – as well as engineering advancement, and now technological innovation. However, few would be satisfied with the steady drumbeat of human loss and harm in driving. Most have removed it from the foreground of their daily lives but would not be comfortable with the continuation of such an aggregated blight in our society. There is a sense that road deaths and injuries are far fewer than in previous times, but they have defied elimination for too long. If the effectiveness of road safety is plateauing, change will be needed. However, safety does not stand alone: it is an attribute of the transportation system. Its tangible assets – policies, products, operations and technologies – certainly help to shape transportation but do not stand alone in creating it.

Transportation has come to be regarded as a "system of systems", a complex system, a leading example of so-called wicked problems requiring boundary-crossing in the application of knowledge and analytics. In addition to complexity, multidisciplinarity is endemic to transportation.

Having materialized in the form of systems engineering in the industry of telecommunications in the mid-20th century, systems thinking has since been applied broadly in social science, political science and medicine. In the case of transportation, the notion of separate modes – road, rail, air etc. – has been used for decades to provide operational focus. More recently, thinking has turned to the potential efficiency gains of multi-modal and intermodal systems, where there is a conscious crossing of existing modal boundaries. Other important boundaries affecting the performance of transportation as a system-of-systems include sectoral distinctions within the industry, the public–private divide and accommodation of both passengers and freight. The further development of transportation practice will require crossing of many of these boundaries. System-of-systems methodologies for analytics and design will need further development, and transportation will undoubtedly provide an early playing field.

The challenges are formidable. Once we include vehicles carrying both people and goods, in modes both private and public, motorized and non-motorized, with personal and commercial motivations, in city, urban, suburban, regional and rural settings, the idea of a transportation system tends to become intractable. That is, the bounds of the system, and therefore its contents and behavior, are difficult to define. When we try to specify a characteristic of system behavior that we care deeply about – like safety – it may be hard to measure and may lack clear means of intervention. And the type of system we are trying to envision may be too limited. For example, a simple engineering system would have external inputs, internal interactions and outputs. The transportation "system" contains a great many engineering systems that are inter-related in some fashion. But transportation also involves many other types of system, including social, environmental, economic, professional and political systems. System-of-systems engineering, thus broadly conceived, is a field that remains very much under development. And transportation is often used by global innovators to test aspects of design and analysis methodology. For example, data generated by personal devices makes multi-modal trips an emerging part of urban mobility.

1.7   COVERAGE AND PERSPECTIVE OF THE PRESENT BOOK

The author aims for an unusually comprehensive coverage in order to trace the evolution of safety knowledge and its application for the benefit of ground transportation. The achievements of the road safety establishment are described through several stages of engineering science and economics, human factors, professional standards, government regulation, disruptive technology and commonly held philosophies.

In the face of a slowing trajectory, road safety is emerging into a new phase, taking better account of growing mechanisms of harm, such as VRUs, and more aware of increasing attention to other important values of surface transportation, including societal support, traffic efficiency, sustainability and social cohesion. Through the adoption of system thinking, we are enabling ourselves to embrace new technology platforms, including advanced driver assistance systems (ADAS), ITS, CVs and infrastructure (V2X) and AVs.

Sometimes it feels to the author that professional concern about road safety is on everybody's lips, but there is a great tendency to sanitize the subject and get away from the violence that still happens out on our roads. The popular author Malcolm Gladwell chose to pepper his early-century New Yorker article on road safety (9) with the exclamation "BOOM". How many have actually witnessed a motor vehicle crash, or even a crash test, and heard that horrific sound? Or seen how long the mayhem lasts until the crashed vehicle finally comes to rest? This book therefore starts with the realism as originally observed by professionals who turned those deep impressions into the evidence needed to develop countermeasures (Chapter 2). Much of this was initiated in the United States and those efforts influenced many other countries, including Australia.

The newly minted enterprise of road safety swung into action and created a generation of safety practitioners, methods, resources and spheres of operation. Again, the United States played a leading role all the way from epidemiology to automotive engineering and testing, government regulation, the science of human factors and the search for crash avoidance. In addition to the United States, safety establishments in Europe, Japan, Australia and Sweden played a major part. Chapter 3 covers this road safety enterprise that became fully operation through the latter part of the 20th century. All three major elements of the road transportation system – vehicles, roads and drivers – were subject to countermeasure development based on extensive study of many crash types. These efforts became more and more dependent on new technological measures contained in vehicles. Again, the author relies on his United States and Australian experience. Note that this period was capped by the articulation of Vision Zero, which was not initiated in the United States but has since had a large strategic influence all over the world.

Chapter 4 moves to describe the evolution of driving itself into the 21st century. This is traced to changes in the transportation system at large, in direct relation to its rising global prominence. In particular, the planet-leading fossil energy consumption of transportation has led a re-evaluation of transportation and greater consideration of its negative societal impacts. In addition to emissions of atmospheric carbon, these include widespread traffic congestion, harmful air pollution and new forms of human injury as walking and biking try to mix with vehicle traffic. Many forms of vehicle restrictions in cities are becoming commonplace along with greater detection and surveillance. Meanwhile, safety research has revealed the magnitude of human lapses, including driver distraction, that cause crashes.

As the norms of driving and moving were being altered, transformational new technologies were made available in road safety. In Chapter 5 the focus is on technologies that assist the driver by operating before a crash occurs and seek to completely avoid the crash or reduce the severity of the crash. These technologies include high-quality wireless communication between vehicles and elements of the infrastructure, driver warnings, advisory driver control actions, crash sensors and automated actions of avoidance. The paradigm of CVs, V2X or C-ITS has developed to the point of providing valuable situation awareness to drivers, along with warnings and alerts for specific crash types. Technological safety intervention, with the human driver still fully in control, is actively addressing the difficult matter of collision avoidance. In addition to the paradigm of V2X, which encompasses V2V and V2I technologies, vehicle-based systems including ADAS and active systems mandated in Federal Motor Vehicle Safety Standards (FMVSS) have emerged in the 21st century. These efforts are underpinned by a pre-crash typology of 37 crash types developed by NHTSA (10). Safety interventions are designed to assist drivers to discharge their prime responsibilities in (1) responding to the demands of the system of roadways; (2) accommodating the actions of others using the road and (3) controlling their vehicle.

Chapter 6 focuses on the prognosis of AVs as a transformational road safety solution. AVs are widely regarded as a future road safety solution of vast importance. It is popularly understood that human error causes the vast majority of crashes and fatalities. Therefore, humans should be replaced by machines that are never distracted, angry or tired. This chapter examines the evidence too so far. Vehicle automation is considered as an emerging safety solution, of a highly technological nature. Also considered is the ability of automation to take up that enormous mantle without creating other safety problems. Several issues are discussed. Firstly, humans cannot be replaced in the transportation system and at least some will continue to drive for many years. Automated driving machines are expected to provide an alternative, but in doing so they introduce a totally new element in the driving system. Secondly, human drivers have a very low failure rate; can it be proven that ADS's will consistently do better? Thirdly, humans provide an enormous value in the daily fabric of driving; will ADSs be able to emulate that? Will ADSs be able to sustain a new modus operandi, a new steady-state version of the dynamic of driving? And if so, how will we navigate the transition from the old status quo to the new situation? Nevertheless, AVs are an irresistible force and may even prove to be the iconic system innovation of the 21st century.

Chapter 7 considers the substantive evolution of road safety from countermeasures to deeper solutions by design. These solutions are enabled by cyber–physical systems' being brought to bear in ground transportation. Solutions are being developed through the paradigms of CVs and infrastructure (or V2X) and AVs. These paradigms are only partly technological: they also entail belief systems, business models and dedicated constituents. Field operational tests and vocational deployments are needed to bring them to fruition. Safety solutions based upon AV and V2X far exceed previous vehicle safety systems because their purview goes beyond the single vehicle (and driver) and includes other vehicles in the vicinity and infrastructure features like intersections. Such technological systems cannot be tested on a conventional test track; and they ultimately rely on a network effect, with most or all vehicles equipped and mutually compatible. Chapter 5 therefore contains descriptions of model deployments of CVs and infrastructure (the so-called V2X paradigm), as well as curated local use of automated driving systems (ADSs) and early commercialization of automotive automation.

Cyber–physical systems must also highly evolved in terms of scale. Their effectiveness depends on a high density of equipped vehicles in the nation's multitudinous traffic streams. This factor of scale is also critical to a quantum improvement in the quality of the nation's driving; in the absence of such technology, human factors know-how is currently missing from two of three crucial responsibilities of driving. The roadway infrastructure is overdue for a technology infusion in order to become a modern cyber–physical system. This is needed in order to support people as better drivers, using benevolent system designs based on human factors principles.

REFERENCES

1. National Highway Traffic Safety Administration (2023) Traffic Safety Facts. 2021 Data. Summary of Motor Vehicle Traffic Crashes. DOT HS 813 515. https://crashstats. nhtsa.dot.gov/Api/Public/ViewPublication/813509.pdf
2. National Safety Council (2023) Injury Facts. Unintentional Injury 2021. https://inju ryfacts.nsc.org/work/work-overview/work-safety-introduction/
3. Buchanan, Colin (1963) *Traffic in Towns: A Study of the Long Term Problems of Traffic in Urban Areas – Reports of the Steering Group and Working Group appointed by the Minister of Transport (Report)*. HMSO.
4. Haddon, William Jr. (1999) The Changing Approach to the Epidemiology, Prevention, and Amelioration of Trauma: The Transition to Approaches Etiologically Rather Than Descriptively Based. *Injury Prevention*, Vol. 5, No. 3, 231–235.
5. The Stanford Encyclopedia of Philosophy. Edward N. Zalta and Uri Nodelman (eds.). Stanford: Stanford University. https://plato.stanford.edu/
6. Mill, John Stuart (1859) On Liberty. Cited in Macleod, Christopher, "John Stuart Mill". *The Stanford Encyclopedia of Philosophy* (Summer 2020 Edition), Edward N. Zalta (ed.).
7. Gerdes, Christian and Thornton, Sarah M. (2015) Implementable ethics for autonomous vehicles. In: *Autonomous Driving 87–102*. Springer.
8. de Tocqueville, Alexis (1841) *Democracy in America*. J. & H.G. Langley.
9. Gladwell, Malcolm (2001) Wrong Turn. *New Yorker*, June 11, 2001.
10. National Highway Traffic Safety Administration (2007) Pre-Crash Scenario Typology for Crash Avoidance Research. DOT-VNTSC-NHTSA-06-02. www.nhtsa.gov/sites/ nhtsa.gov/files/pre-crash_scenario_typology-final_pdf_version_5-2-07.pdf

2 Sixty Years of Motor Vehicle Harm

2.1 MID-CENTURY EPIDEMIC OF SEVERE INJURIES INSIDE VEHICLES

The post-war world of the 1950s and 1960s became increasingly aware of the injuries and fatalities caused by motor vehicle crashes in everyday driving on all types of roadways. This didn't just occur at particular times, or in particular locations, but serious crashes – sometimes horrendous – seemed to be almost endemic to the act of driving. Post-war optimism and growth was being tempered by the realization that there may well be a price to pay for the newly created freedoms of driving nicer cars longer distances, and living far from the mundane surroundings of places of work – offices and factories. It is telling that the medical professionals, those in the emergency rooms and operating theaters, were the ones who sounded the alarm.

2.1.1 Escalation of Blunt Trauma in Emergency Rooms

Most fatalities and severe injuries that occur in collisions of motor vehicles are caused by blunt trauma. Severe internal injuries are caused by blunt forces imposed by hard vehicle surfaces, or by high rates of deceleration. Forces cause direct damage to internal organs and deceleration causes separations between organs. This type of trauma commonly occurs to the head, chest and abdomen. In addition to the brain and the heart, affected organs include the liver, spleen, colon and bowel. Traumatic brain injury is a common cause of fatality in motor vehicle crashes. Blunt trauma is difficult to diagnose and repair, and even small delays in treatment increase the risk of fatality. Though less common than blunt trauma, penetrative trauma also occurs in car crashes and was more common in the early days prior to more protective design of car interiors.

The actual injuries are many and varied, and they are extremely confronting. It is perhaps unsurprising that the violence of many crashes tends to be sanitized through the use of professional jargon, and through the surrogacy of crash types. For example, crash databases generally identify crash severity through the outcomes of loss of life, several levels of injury ratings and property damage only. Loss of life is even parsed according to how soon death occurs. Crash description tends to begin with crash

DOI: 10.1201/9781003483861-2

type: rollover, head-on, rear-end, side-swipe etc. Road safety generally avoids the use of pejorative terms like dangerous, or violent, and prefers terms like life-threatening, or "requiring hospitalization". It has long been believed that graphic reality does not encourage safer driving and may even tend to block out safety messaging in people's minds. Therefore, the emotional and psychological impact of the human harm seen directly by physicians is not conveyed to the general population.

The world continued to turn. New and glossy ways of selling cars arose, including auto shows that provided a regular cadence for the launch of new brands and models, some of which offered non-standard options and capabilities. It became much easier to covet certain types of cars and to develop a knowledgeable appreciation of favorites. The concepts of economy, reliability, luxury, status, pride of ownership and brand loyalty were invented and refined over the many years of success enjoyed by a highly capable automotive industry. In the United States, the General Motors Corporation became a bellwether for the state of the national economy. Other countries, including Japan and Germany, also placed their auto industries at the very pinnacle of national importance.

2.1.2 ATTENTION ON THE AUTO INDUSTRY

While there was definitely a convergence on the part of car designers around the visual cues of aerodynamics, in every other way there was an explosion of diversity: size, shape, vocation, lifestyle statements and the like. The appearance and reality of speed was ever-present and a kind of cult arose that took particular account of the driver. The driving position – now intended for more than the basic control of the vehicle – was comfortable and even could be described as elegant and rewarding. Various aspects of vehicle performance were quantified and even lionized. The mysterious notions of vehicle handling and ride quality were expertly reviewed and appreciated by connoisseurs all over the world. At the same time, the popularity of motor racing prompted the appearance of sports cars intended for everyday road use, and vehicles with sports performance settings.

In 1965, the consumer advocate Ralph Nader published a best-selling non-fiction book (1) about automotive engineers' disregard for safety, entitled *Unsafe at Any Speed: the Designed-In Dangers of the American Automobile*. A well-known trigger for the book was a hidden propensity of the early-1960s GM Corvair model to roll over. Ironically, this car was marketed as a vehicle with a definite sporting image. The book tied the Corvair's alarming form of vehicle handling to rear suspension design. I note that expert opinion at the time was divided on such a strong cause-and-effect statement. But Nader definitely didn't stop there. Nader also covered a range of other issues – a mix of what we would now call crash avoidance and occupant protection factors. At the age of 31 years, Nader had produced an expose that literally changed the world. His fundamental shift in thinking was that injuries were caused by the "second collision", between the occupant and the car's interior mechanics and surfaces. At that point, he didn't think much could be done about the "first collision", with another vehicle or a roadside object, because human drivers could be so unreliable.

The implications of Nader's book went beyond engineering priorities to the motives of large automakers, and in some cases to their ethics. According to Nader, the industry had for many years cast safety in the form of Engineering, Education and Enforcement. In his view, the auto industry portrayed the latter two Es as references to the responsibilities of the driver, with the first E referring to the designers of the roadway. In other words, responsibility for safety was being pushed away from the vehicle and its manufacturers.

Nader was not alone in exposing road safety and taking it from a personal matter to a marquee cause for the media, and soon attracting congressional attention. Ambitious senators and congressmen started to compel the large, rich automakers to answer questions at congressional hearings. In the event, the industry's response and approach to crisis management left a lot to be desired. In Elizabeth Drew's wonderful Atlantic article of October 1966 (2), the political machinations leading to attention-grabbing public hearings are described in a perfectly acerbic way. She characterized Nader as a "basically sound fanatic" who was hard to counter with logical argument. Important political leaders were swayed. In 1965, no less than Senator Robert Kennedy skewered the President of General Motors with "you made $1.7 billion last year... .and you spent $1 million on this (safety)?". Then, in the now-classic manner of the cover-up being worse than the crime, GM hired private detectives to provide a rundown on Ralph Nader the man, and "what makes him tick". When this was revealed, there was a degree of shock that America's finest corporation could stoop so low. Congress acted quickly and without compromise. Nader became the singular trigger for vehicle safety rules that were no longer going to be voluntary – they were going to be mandatory. And not in the fullness of time, but next year and the year after.

The truth is that, at the time of Nader's singular activism, severe injuries were being caused to unrestrained vehicle occupants by fixed steering columns, hard and rigid interior surfaces, poorly fixed seats, weak door and roof structures and the propensity to be thrown through the windshields that shattered and fell apart. While human occupants fared very poorly when they collided with interior surfaces of the vehicle, they were even worse off when thrown from the vehicle. The auto industry had long maintained that all crashes are caused by drivers, but the violence occurring in and around their most highly engineered vehicles after they rolled off the production line was hard to shake off. Their intemperate approach to Ralph Nader and other activists tended to drown out their more professional responses. Drew quotes a senator who said, "It was that Nader thing. Everybody was so outraged that a great corporation was out to clobber a guy because he wrote critically about them. At that point, everybody said the hell with them".

Far from Washington, DC and Detroit, the attention of the Australian State of Victoria was galvanized by the Royal Australasian College of Surgeons. Emergency room physicians were taking the time to publicly expose the road toll, including the life-long trauma visited upon the survivors. They were seeing hospital wards populated with an increasing stream of irreversible head injuries and spinal ablation. At the same time, deaths resulting from such crashes reached an all-time Australian high in 1970. Activist academicians pursuing the new field of human engineering joined with legislators to bring about the world's first compulsory seat belt law – enacted in Victoria in 1970.

2.1.3 DRIVE FOR MOTOR VEHICLE SAFETY REGULATION

As a direct result of Nader's political impact, the federal government created the U.S. Department of Transportation in 1966, and then the National Highway Traffic Safety Administration (NHTSA). NHTSA was granted powers to require regulated entities – specifically automakers – to comply with newly formulated vehicle safety standards. These standards were called Federal Motor Vehicle Safety Standards (FMVSS) and eventually covered many specific aspects of occupant protection, as well as crash avoidance.

The first director of NHTSA was the epidemiologist William Haddon who focused on the injuries that were caused by the second collision. In his 2001 *The New Yorker* article entitled "Wrong Turn" (3), the author Malcolm Gladwell attributed the following words to Daniel Patrick Moynihan: "He never forgot that what we were talking about were children with their heads smashed and broken bodies and dead people". He was referring to his protégé William Haddon. But very few managed to keep the human harm in their sights without professionalizing it.

Together with his patron Daniel Patrick Moynihan, William Haddon promoted the idea that motor vehicle injuries represented an epidemic. Throughout the 1950s and 1960s vehicle fatalities increased at their most rapid rate. Public administrators were worried that the human toll would continue to increase, virtually out of control. Whether or not motor vehicle injuries actually represented an epidemic, Haddon's use of epidemiology turned out to be a highly effective methodology and was a propitious starting point for his new federal safety regulator, NHTSA.

2.2 CREATION OF THE U.S. NATIONAL TRAFFIC SAFETY ADMINISTRATION (NHTSA) AND MOTOR VEHICLE SAFETY REGULATION

In 1966, President Lyndon B. Johnson appointed William Haddon as head of NHTSA. As a physician and epidemiologist, he adopted a public health template for the development of a comprehensive range of countermeasures. His famous Haddon Matrix required that such measures were actively sought in each of the domains – driver, vehicle and environment – as well as their relevance before, during and after the crash. Haddon used the analogy of eradicating malarial mosquitos from swamps. The injured party, the driver, was considered to be the "host" and the interior of the crash vehicle was the "vector" of harm. Total harm was considered to arise from combining drivers' exposure in the road system, risk of particular types of crashes and the consequences of crashes. NHTSA has by now pursued its regulatory mission with great diligence for more than 60 years and has drawn on the highest levels of expertise in safety science, automotive engineering and human factors research. But it all started with coming to grips with the vector and its toll on human bodies.

NHTSA was granted the power to develop and mandate motor vehicle safety standards. Such standards were designed to counter the types of injuries that were being observed on a regular basis. The earlier years were devoted more to vehicle occupant protection measures in crashes. In later years, the focus moved to means of avoiding crashes through countermeasures development in the pre-crash zone of the matrix that has served as NHTSA's north star throughout. In 2007, NHTSA (4) recharged its methodical process of removing crash scenarios from the table by painstakingly identifying 37 pre-crash scenarios. Each scenario in this typology is defined in sufficient detail that new driver assistance technologies could be brought to bear in an orderly manner.

The administration of such standards required new research into crashes, means of mitigation and methods of evaluation. The development of the will, and the means, to regulate vehicle safety may be viewed as world-changing and it continues to this day. Many countries developed similar capabilities, and there is now a range of global bureaucracies trying to keep such widely applied standards active, and in reasonable alignment, while remaining conducive to international automotive trade.

These actions were by no means silver bullets, but U.S. traffic fatality rates – at an all-time high in the period 1966–1972 – reduced inexorably over the ensuing decades. Taken over the now hundred years' history of crash data, fatality rates have reduced 92%. We will see that such rates, expressed as fatalities per mile or km of travel, or fatalities per unit population, only tell part of the story. While technocrats and many others may consider this a sign of great success – in the sense of the "transportation system" – the total number of fatalities did not decrease and has hovered around 40,000 ever since, having peaked over 50,000 deaths in the early 1970s. As the rates decreased impressively, increases in population, annual miles driven and urban intensity counteracted such improvements. So complete success at the societal level has continued to elude us.

As the new U.S. DOT set about laws that required certain safety features on vehicles, they also needed a means of ensuring that such measures provided benefits greater than their costs. Thus, the enduring practice of benefit–cost analysis in vehicle safety was born. This required consensus concerning analytical methods that combined engineering thinking with the real-world rationality of economists. We have now become accustomed to the idea that the face value of a given safety measure probably needs to be discounted in practice. Inevitably, other important things change when the singular measure is introduced. For example, buyers of the new vehicles may tend to drive a little faster when the vehicle is safer in some objective way.

By 1975, leading economists of the Chicago school had turned their attention to vehicle safety regulation. Sam Peltzman (5) applied economic principles to NHTSA's inaugural round of mandatory requirements, which contained the following safety countermeasures:

- Seat belts for all occupants;
- Energy-absorbing steering columns;
- Penetration-resistant windshields;
- Dual braking systems; and
- Padded instrument panels.

Note that the majority of these measures were intended to negate jagged vectors that directly penetrated and battered human bodies. It was much later when more arm's length "safety-related" measures came into play and safety professionalism began to rule the day.

2.2.1 THE UNDERBELLY OF CAR REGULATIONS – DRIVING INTENSITY

When the notion of "driving intensity" was introduced, Peltzman's economic modeling showed that the above safety rules may have no impact in reducing fatalities; according to Peltzman fatalities could even increase. Driving intensity implied an amalgam of risk taking either at the discretion of how aggressively the driver was choosing to drive or through the increased presence of riskier drivers, such as younger drivers. The idea was that risky things were waiting to happen, once it was made known that the risk purely related to the vehicle was reduced. In a classic supply-and-demand analysis, economists saw driving intensity as a good with a certain degree of attractiveness to the consumer, just as safety was a desirable, but competitive, good from the consumer's perspective. Fortunately, NHTSA and the automotive engineers won the day. It is extraordinary to look back at the five measures above and see their absence as a consumer good. Seat belts alone have saved several hundred thousand American lives.

Driving intensity was not defined in these early moves to regulate vehicle safety. Clearly the significance of a safer vehicle could differ dramatically between individual consumers. A consumer's perception of the relative safety of their vehicle, and the ways in which this would interact with all of their other daily mental processes, was a hazy idea then and would largely remain so. Considered as a consumer good, the weight they would place on the inherent safety of their vehicle – and how it would vary over both short and long time scales – has barely been contemplated in a scientific sense. And we have seen total road crash fatalities remain stubbornly constant over a period of a half-century – proving the existence of strong factors counter to inherent vehicle safety. Some would simply call this a rebound effect.

2.2.2 NHTSA's TEMPLATE FOR VEHICLE SAFETY REGULATION

Once the NHTSA regulatory template was established, vehicle safety regulation had settled into a symbiotic relationship between industry and government. Both sides used a wide array of expertise, including engineers, economists and lawyers. When NHTSA receives criticism, it comes from both sides of the tug-of-war for more or less standards. Public safety organizations generally see a need for tougher requirements, while automotive industry organizations are opposed to excessive regulatory zeal on the part of NHTSA. A balance was struck and the half-century between the mid-1960s and the mid-teens proceeded to be the age of vehicle safety regulation. NHTSA's evidentiary process and economic analysis has stood the test of time. It represents a special case of government regulation as practiced broadly by economists throughout the 20th century – sometimes known as the regulatory age. Regulation writ large was applied to markets in very large and essential industries such as energy, telecommunications, banking, finance, food and drugs, and many more. By comparison, NHTSA's style of regulation was quite specific.

Philosophers like John Stuart Mill (6), writing almost two centuries ago, saw the vast potential for government interference and resulting loss of liberty. He particularly prized continuous improvement – following the "progressive principle". He believed that the government should only exercise a preventive function if "mischief is certain". The emergency room physicians and the economists of 50 years ago held very different opinions on the certainty of harm in crashes. Road safety has succeeded in convincing most of us that evidence of risks, together with relevant preventative measures at reasonable cost, is sufficient to justify government interference.

Was there an alternative to vehicle safety regulation? Early thinking was that litigation was a direct alternative to regulation. Regulation was viewed as a good like any other, subject to the laws of supply and demand. This notion worked in some contexts. For example, if the rail company had a state monopoly on hauling lumber, a trucking company could sue for the right to enter the market. The rail company could lobby state legislators for regulation of lumber to rail; this would probably be preferred by the government. However, all parties would be sensitive to the criticism of "regulatory capture", and so the regulatory good of lumber carried solely on rail would need to be rationed. Strangely enough, mid-century economists often cut their teeth on freight regulation, where issues were relatively straightforward and data publicly available. This was not the case with road safety. NHTSA was not interested in economic regulation – it pioneered safety regulation.

Government intervention to make each and every vehicle safer – from the moment it rolled off the production line – transformed road safety forever but never made total sense to economists. They understood that the consumer presented with the object of a safe vehicle, and operating in an automotive world containing other advancements, may no longer be the same consumer. And the rush to more vehicles, driven further and faster by more drivers, to more destinations, for more seductive purposes, on more conducive roadways, could have been a safety disaster. This was totally averted. In the parlance of the economist, vehicle safety regulation was probably a misnomer. What we have is vehicle injuriousness reduction through government intervention, at reasonable cost.

The ethic driving Haddon's apostles of public health was harm reduction. There is no doubt that the moral clarity of this effort allowed it to happen on a national scale and eventually a global scale. Meanwhile, there was always a strong strain of egalitarianism in the fact that each vehicle sold to consumers met certain standards of safety design, and each consumer was free to choose any vehicle, within reason. And the system had a flip side of personal freedom in that driving intensity was not regulated. This allowed for the burgeoning societal and economic benefits of mobility to flow throughout a half-century, and continuing. As we will see, harm reduction is no longer a clarion call, and the rate of safety improvement has slowed.

2.3   RESEARCH INTO MOTOR VEHICLE HARM, DRIVING SCENARIOS AND DRIVERS

As soon as the long journey of harm reduction was commenced in the mid-1960s, research was initiated to fill in the many gaps concerning the driver's influence on safety. The new field of human engineering – closely related to the science of human factors – sought the input of applied psychology and aspired to understand man–machine systems. Safety research was therefore interdisciplinary from the beginning. How could you improve the driver's ability to respond to a threat moving at high velocity without understanding something about the human ability to comprehend, assess and act under such situations? The speed with which such processes could be dealt with was only the most obvious of the matters under consideration. And how could one begin to design a less injurious interior of the vehicle without medically based knowledge of the impact resistance of humans and internal mechanisms of serious injury? The enormous thirst for information led to extensive applied research programs housed in universities, government agencies and companies all over the world.

2.3.1 HUMAN IMPACT TOLERANCE

Research universities such as the University of Michigan operate hospitals that have access to large-scale data on serious injury in motor vehicle crashes, as well as cadavers for experimental purposes. Biomechanical testing of human remains continued for many years until crash test dummies and computer simulations developed sufficient fidelity to take over. The physical limits of humans in their almost infinite variety are now well established, as are the optimum ways to manage crash energy and limit its transfer to vehicle occupants.

At the start there were no crash test facilities, or sleds for testing human limits of deceleration. Mid-century, this began to change. In 1954 Colonel John Stapp (7) of the U.S. Air Force had personally endured a massive deceleration of 46 g, riding a decelerator sled. Subsequently he was loaned to NHTSA in 1967 to work on the survivability of car crashes. Some of the first tests of human resistance to high impact were carried out by resolute scientists like Albert King at Wayne State University in Detroit, who simply dropped instrumented cadavers down elevator shafts. The horrors of motor vehicle injuries that had to be confronted by surgeons in emergency rooms justified the macabre experience of those who were prepared to carry out this type of testing. The trauma of someone arriving at the hospital with a fence post still protruding through their body extended beyond the crash victim to the whole emergency team who worked desperately to save them.

2.3.2 PURPOSE-BUILT FACILITIES AND METHODOLOGIES

There was a deep-seated motivation to reduce and avoid such trauma. The sense of urgency was palpable and much progress was made in managing the immediate human consequences of crashes. However, the hopes, fears, passions, motivations, situation awareness and decisions of human drivers as avoiders of crashes are another matter. The early hopes for useful man–machine integration were not fully realized. Only very limited aspects of human sensory, cognitive and motor performance may be simulated. Even the most advanced driving simulators are not relied upon in the fundamental safety design of vehicles. In latter years, a driver assistance application such as lane departure warning would be much better assessed in a naturalistic driving study. This meant a relatively large number of real drivers, subject to extensive observation and data collection, over a long period in many and varied driving environments.

A short time spent observing the passive face video from naturalistic driving tests will dispel any notions of trying to isolate driving as a defined task subject to universal improvement. The background data streams – including distance to other vehicles, movements within the traffic stream and responses to traffic controls – show the extreme variability between drivers. Whether such variations are caused by state of arousal, impairment, attention, distraction, perception, information processing, driving skills or driving ethics, it is impossible to know.

None of the detailed driving data obtained over many years has shed much light on driver appetites that may counteract the improved safety design of vehicles. The subject of "driving intensity" has therefore remained well under the radar and is little better defined than it was 50 years ago. The fallback assumption seems to be that a combination of more drivers, vulnerable persons on the roadway, more miles, including driving on lesser roadways, has worked against hard-won vehicle safety improvements.

2.3.3   EVIDENCE OF WIDESPREAD POOR DRIVING

But we should be aware that several unpalatable aspects of driving have come to light over the years. The idea that drivers would fall asleep while behind the wheel, and cause a fatal crash, came as a surprise to many. However, naturalistic face video confirms that drivers of both passenger cars and freight trucks fall asleep, and this happens on both long and short trips. This problem is more difficult to detect by police when they investigate roadway crashes. Professional drivers, and no doubt personal drivers, are affronted when suspected of falling asleep and causing a crash.

Another unexpected example from the black box of driver behavior is diversion of attention, usually termed distraction. The addictive nature of personal devices has proven too much for many drivers who seem to be willing to relegate driving to the status of a background task. They also carry out a lot of culinary and grooming activities. And video data from naturalistic driving studies – even when drivers were well aware of being observed – reveals some astounding acts carried out in moving vehicles. It certainly seems that driving intensity – our shadowy competitor to design safety in vehicles – is not entirely a consumer good. It is probably not a continuum like miles traveled and it does contain harmful tendencies like falling asleep while driving and striking pedestrians.

One does not need to be a card-carrying libertarian to acknowledge the value of increased usage of vehicles that are designed and certified to be safer. The wild card is the driver, and scientific research has struggled to find ways to improve driving behavior. Even among human subjects who agree to be part of a naturalistic driving study, there is little patience for imperfect devices intended to recognize risk and warn the driver. So-called false positives – when a warning is given by mistake and the risk is not present – will quickly lead many drivers to switch the thing off. Some drivers have a short fuse when it comes to warning applications that may demonstrably be in error. Such decisions are often taken lightly because most consumers are ill-informed about the safety benefits of newer, reasonably priced driver assistance systems that will improve their driving performance.

There is a general reluctance to condemn the tendency of drivers to allow themselves to be distracted, or to drive when tired. Safety researchers have had much trouble even getting the terminology right: are they fatigued, drowsy or just plain asleep? Once we start getting into the psyche of the driver – and get away from our professional assumptions about a semi-automaton carrying out an understandable and reasonable task – we can find some black-hearted surprises. The advent of road rage concerns reasonable people. But who knows the range of private – perhaps selfish – thoughts that could influence driving behavior? Am I driving a large sport utility vehicle (SUV) so that I will be better off in a vehicle-to-vehicle collision with a lighter, smaller vehicle? Am I quietly putting my thumb on the scale to protect my family? And what of the driver of a large, stiff and heavy long-distance freight truck who won't risk avoidance action when confronted with a car traveling toward them on course for a head-on collision? Such decisions are beginning to surface with machine-driven cars – will they be programmed in an ethical manner, and by whom? But we should not pretend that the humans who are currently reacting on the spur of the moment are always making considered, ethical decisions.

Mid-century economists certainly started something in insisting that there is more to safety than straight ahead harm reduction. But are we really talking about two competing goods, namely, safe vehicle design and the pursuit of driving? Unfortunately there is a great deal of forbearance in the way we treat poor driver behavior, and the pursuit of driving is not all good. If we pick this apart we may find strong evidence of an egalitarian, and perhaps libertarian, society at work. Even when there is sufficient evidence after the event for police to bring charges, courts are notoriously reluctant to impose harsh penalties. Is it the fleeting, banal nature of the offense, or the circumstantial nature of the case? With our attention riveted for so long on the consequences of the crash, it is not surprising that driver intentions remain largely unknown. What seems to be lacking is a willingness to assume responsibility – what is my reasonable responsibility toward others on the road? How can this be brought into the light of day?

2.3.4 BROAD AND HARMFUL MIND-SET AMONG DRIVERS

Motor vehicle safety regulation is not a good that is sought by the regulated entity, the automotive industry. It is therefore a misnomer, perhaps introduced in order to make it more palatable in its early years. Vehicle safety is government intervention to prevent harm, a modern version of Mill's official interference to prevent certain mayhem. That mayhem is apparent to all concerned in the road transportation system. There is no question that intervention is needed. What of the competing good – driving? The good of driving also has its dark side through the lack of responsibility demonstrated by many of those who drive. Until all of these harmful tendencies are understood and removed by interventions under our prevailing social contract, economists are not in a position to model safety regulation in its true sense. We also need to be very aware of layers of abstraction obscuring the harm. Nobody really wants to write about people being beheaded in collisions. Though bloodless, the broad tendency of irresponsible people getting behind the wheel and injecting unnecessary risk into our transportation system is also an unpleasant subject. Nevertheless, the unvarnished harm is what motivates and centers road safety.

Road safety's observation of driving harm has been overwhelmingly tied to bodily injury. However, we should also view wanton lack of responsibility of drivers as a form of harm in society. We are comfortable with the utilitarian approach of road safety because it starts with the harm embodied in human injury. The mental harm of irresponsible drivers surely requires a different philosophical frame invoking ethics and morality but is just as real. While physical harm in collisions was apparent from the very start of the modern road safety era, mental harm such as distraction was diagnosed much later through technologically based study of crash causation.

2.4 HOW BIG AND BAD IS THE ROAD TRAFFIC INJURY PROBLEM, AFTER ALL THESE YEARS OF INTERVENTION?

How do vehicle collisions compare with other sources of preventable death? The National Safety Council (NSC) reported the top three categories of preventable death in 2021 to be poisoning, motor vehicles and falls (8). Viewed in terms of lifetime odds of being killed, motor vehicle crashes equate to suicide and gunshots and are less risky than opioid over-doses. Motor vehicle crashes are also less dangerous (odds reduce by a factor of 5) than typical pedestrian activity or motorcycle use. We have a 1 in 6 chance of dying from heart disease, 1 in 58 from opioids, 1 in 93 from motor vehicles and 1 in 1,300 from fire or smoke. The odds of death in a motor vehicle are far from minimal because people spend a lot of time driving, and this applies right through their lives. We only need to "follow the money" to appreciate how these sorts of odds are baked into our society: motor vehicles represent one of the largest forms of household expenditure, along with housing.

Road fatalities are as big a problem today as they were in the mid-20th century. That problem is given extra weight by the fact that younger cohorts of citizens are as badly affected as older cohorts, who naturally experience many more deaths from disease. So it is common to give close consideration to the extent and trend of the crash problem for the age group up to 40 years of age. In the United States, the total number of younger persons killed in road vehicle crashes – including drivers, passengers, pedestrians, cyclists and other road users – has hovered around 40,000 for that entire period and has tended to reduce. What has changed is that the new century brought many deaths from drugs in the younger cohort. In 2000, such deaths were a small fraction of motor vehicle deaths, but today drug deaths are about double unintentional deaths in vehicle crashes. This should not be interpreted as evidence of success with road safety.

2.4.1 SO MUCH DRIVING

Motor vehicle deaths have not been eliminated, or close to eliminated, relative to other causes of unintentional and intentional mortality. On the surface, there have been no dramatic changes. It is surely of concern that traveling in road vehicles still causes deaths in numbers comparable to homicide and suicide. Even though vehicles have become much safer, the sheer volume of driving that occurs each year has increased dramatically. This is caused by population increase, along with increased vehicle ownership per household, and many other things in life. For example, longer commutes, and the easy access to more distant attractions also increases miles driven. So much so that total miles driven have increased four to five times since the 1960s. The nation's drivers now cover more than 3 trillion miles each and every year.

Irrespective of the hopes and dreams of vehicle owners and drivers, the mid-century economists' notion of driving intensity became a massive reality. It was able to quietly multiply itself several times over while crashes did not change very much. So if the purpose of the road system is to support dramatic growth in miles driven – for whatever reasons – the road safety community has had a certain degree of success.

2.4.2 IMPROVEMENT IN THE CRASH RATE

What changes led to such a marked improvement in the crash rate? Among safety experts and researchers, it became customary to think in terms of an interactive system of three distinct elements: the vehicle, the driver and the environment. Obviously safety improvements flowed from long-term developments in each of these elements. It would be difficult to separate these contributions in a quantitative manner. Nevertheless, the safety improvements within vehicles – both to protect occupants when crashes do occur and eventually to avoid crashes – would stand alone in the application of advanced technology and the speed at which change could actually occur. The era of improvement in the crash rate was curated by a significant federal effort for informed and professional intervention in the service of road safety. Before the 1960s, this did not exist. Following the publication of Nader's *Unsafe at Any Speed*, the U.S. Department of Transportation was established in 1965 by upgrading the Federal Aviation Administration to a cabinet post and folding in other modal administrations. The first of these was the NHTSA, and this was joined by the Federal Highway Administration (FHWA) and the Federal Motor Carrier Safety Administration (FMCSA), among others. Like NHTSA, the latter two new administrations also had a strong safety focus.

The institutional capabilities in road safety did not stop there. The National Transportation Safety Board, with a scope covering all modes, was established in 1967. The Volpe Center, designed to provide deeper analytical expertise across all parts of the DOT, was established in 1970 and has long served as the final arbiter of the value of new vehicle safety regulations. The NSC is a prominent non-profit member organization established in 1913 and covering health and safety across a range of sectors in the economy. The American Automobile Association (AAA) Foundation for Traffic Safety has been dedicated to highway safety research and education since its creation in 1947. The Insurance Institute for Highway Safety, established in 1959, is dedicated to safety evaluation of vehicles and pioneered laboratory crash testing of whole vehicles. In 1979, NHTSA commenced the New Car Assessment Program (NCAP), in order to issue a standardized safety assessment covering each and every vehicle on the U.S. market. And safety laws at the state and federal level have been pushed since 1989 by the broad-based Advocates for Highway and Auto Safety.

From the paltry $1 million ascribed to being the sum of General Motors' safety effort in 1967, mighty efforts of technology, manufacturing, construction and government intervention have materialized. In the absence of Ralph Nader and Robert Kennedy, how did this get going and keep going? It would not have happened without the great American propensity to associate. As viewed in its beginnings by Alexis de Tocqueville in the earliest days of the Republic, associations greatly influenced the lawmakers. Otherwise we would have risked the tyranny of the majority, mediocracy and stagnation. Associations take many forms. In this case, each of the industry sector associations across the road safety spectrum took on the question of "what does road safety mean for us?" As the 20th century entered its latter decades, such associations were extremely well equipped, given the early-century age of professionalism – especially in engineering – and the late-century era of regulation. In the case of road safety, industry associations needed to engage and push back on the new propensity for government intervention.

Industries had learned to pick their fights very carefully when it came to opposing regulations. The auto industry was the main industry target, and they had been greatly chastened by their experience in the mid-1960s. Even though roads were largely built and operated by a multitude of government agencies, they were the lumbering, benign giant of the evolving road safety enterprise. While highway traffic standards – including safety standards – were adopted by road agencies, such efforts were not legislated, and they are distributed across the 50 states.

2.4.3 NHTSA's Evolving Game Plan

NHTSA continued on its path of formulating and administering compulsory vehicle safety standards. Over the years, the pace of new standards slowed as NHTSA's focus moved from protecting people in crashes to avoiding crashes in the first place. Safe design and safety features became a permanent part of the businesses of automakers and their suppliers. The industry maintained that vehicle safety was "pre-competitive" – safety was not considered to be a major factor in selling more cars. This was not always the case. Some U.S. manufacturers believed that certain European manufacturers had a design philosophy of reducing visible crumple when one of their products crashed. This may give the impression of crash resistance to consumers (even though greater absorption of the energy of impact is known to be safer). Safety became a subtle presence in automobile marketing as it became a common perception that luxury vehicles were safer than entry-level models. Sales of mid-level and luxury vehicles also involved optional offerings of bundled avoidance technologies, at escalating price points.

In 2019, NHTSA determined the cost of U.S. road crashes to be $340 billion (9). In absolute terms, this is a high cost; for example, it is comparable to the national cost of storm and flood damage. Meanwhile, federal public expenditure on highways – for any reason, including safety – is contained within a highway bill of around $200 billion. Given that safety is only a fraction of this total, it is clear that the annual cost of crashes far outweighs public expenditure on crashes. Total consumer expenditure on vehicles sits around $700 billion. Even if we assume that 10% of this total is targeting safety, this amounts to about $70 billion sheeted home to the consumer. So the total spent on public and private road safety is still only a fraction – probably less than half – the cost of road crashes. That cost is a many-tentacled beast, including years of life lost or impaired, hospital and rehab, vehicle loss and repair, insurance, emergency services etc.

Whether the cost is counted in human loss or dollars, injurious roadway crashes still impose a nationwide problem that looms larger than it should, given the expertise and energy that have been devoted to it. Road safety is not a system in itself but is an attribute of something much bigger and more concrete – road transportation. And road transportation is not entirely a system in the sense of operations, administration and economic regulation. It is not a fully professional system, or a closed system. Professional management and oversight of road transportation, and its safety, cannot be successful until its endemic human harm – in forms both exterior and interior to the driver – are brought under control.

2.5   THE DELICATE SUBJECT OF "CRASH CAUSATION"

Drivers cause all crashes – right? At least that was the statement made by the head of General Motors in 1967. Safety experts later pivoted to the position that crashes have no single cause, so the accepted terminology is that several factors, or even a multitude, contribute to the occurrence of a crash. These factors could include aspects of vehicle and roadway, as well as the ubiquitous driver. Note that crashes used to be called accidents, but the nihilistic tone of this term was found to be unproductive – crashes are not random, blameless events. They actually have causes – well, they have contributing factors! Straight away we will note the lack of clarity creeping in. Why this unwillingness to call a spade a spade? Perhaps we had a desire to keep the whole "system" – the vehicle, the roadway and the environment – in the frame of crash causation, as well as the driver.

Those of us who have investigated individual, specific crashes know that each crash has its own story. Many extraordinary things – unbelievable things – happen until the dam of good fortune breaks and the crash occurs. There is no inevitable pathway, no typical chain of events, but the dam eventually just breaks inside a driver's head. Often we are puzzled when there is no evidence of evasive action, but the dam had already broken. An example from my own experience: a learner driver is driving a major highway under the supervision of his father. Even though it's a major highway, it is undivided. His father happens to be very controlling – a martinet. While the open road speed limit applies, there is a special rule at the state level that requires a learner driver to never exceed a much lower limit. The father is constantly making sure that his son obeys the special state rule. A long-distance truck rapidly approaches the student's car from behind and impacts the rear of the car. The impact is only moderate, but enough to bend the car and jam the driver in place. The friction when the car is pushed along the road surface is enough to set the car on fire. The father is able to exit the car and is helpless watching his otherwise-uninjured son being incinerated. There is no evidence that the son was aware of, or reacted to, the impending crash.

No categorization could adequately capture this crash from a prevention point of view. The list of contributing factors is very long. Some of these factors simply reflect the prevailing status quo, some are choices on a timeline well upstream of the crash and some are rapid-fire happenings immediately before the crash. The many crash factors also apply to all participants: the car and the truck, the two drivers and the car passenger. And who or what is the cause of the crash? Even if we agree that the car should not have been traveling so far below the general speed limit, who or what caused this crash? Even if we agree that the father should never have chosen this highway for a learner driver, police investigators tend to look for fault much closer to the moment of the crash. If we learned that the truck driver had been texting, then the case is closed: in the parlance of police investigators, the truck would have become "Vehicle A" instead of "Vehicle B". And Vehicle A always absorbs the lion's share of crash-investigatory effort.

So causation can indeed be a fool's errand. But the long list of contributing factors also confounds effective countermeasures. Even if we somehow fixed some, or even most, of these factors, how long would it be before they – in combination – conspired in another bad crash? Multi-factor crashes of this precise composition may be very few and far between, so that the overall impact on crashes could be minimal. Otherwise, we simply don't know how important each of the contributing factors is individually. So we could not proceed to countermeasures with any real confidence.

2.5.1 TWO CRASHES

Ralph Nader's original contention was that each crash was actually two crashes. Firstly, a vehicle striking another vehicle or a fixed object like a tree. Secondly, the unrestrained occupant striking the interior surfaces of the vehicle. Clearly the first crash hinges on the ability of the vehicle structure to absorb energy and retain interior space for survival. The second crash depends a lot on the occupant restraints, such as seat belts and airbags, and the presence of protective surfaces inside the vehicle.

But there may be little point in further considering the cause of these two crashes because naturalistic studies show that they are usually preceded by a sudden driver failure. Not exactly a crash, but more of a tragedy in the true sense of the word – a very sad event caused by a dramatic flaw in the protagonist. The original word "tragedy" also included a meaning of grinding inevitability – that continually pursuing the same systematically flawed types of actions would result in downfall.

When parsed this way the crash as we know it is part deterministic – Nader's two crashes – and part mental failure. The mental failure is in turn partly predictable – subject to alcohol, fatigue or distraction – and also unpredictable. We just don't know what unrelated thought streams may suddenly be overwhelming the driver's faculties – those faculties most desperately needed for the avoidance of harm. Willful failures include speeding and intersection violation. Casual negligence includes texting, reading, writing, reaching for objects inside the car and extended glances outside the vehicle.

2.5.2 The Harmful Propensity for Human Lapses while Driving

Close examination of the naturalistic crash event data from the Transportation Research Board's Strategic Highway Research Program 2 (SHRP2) (10) supports the idea that pervasive low-level distraction is so widespread as to be a fact of life, rather than a specific act on the part of the driver. Then much more dangerous failures are overlaid with reasonable frequency. These failures are temporal and immediately put the vehicle in harm's way. They are usually either deliberate or irresponsible. The emergence of such human failures on a large scale is itself a form of harm to the fabric of driving and covers a wide range: distraction, unresponsiveness, negligence, recklessness and aggression.

By allowing the accumulation of such a veneer of distraction, the driving population is acting on their perception – one based in reality – that crashes are a rare event from an individual's point of view. This is clearly a widely believed perception. Deliberate and irresponsible personal failures are definitely causing crashes. If we view each of these as a unique tragedy, then the flawed protagonist question arises. Should flawed individuals be at complete liberty to drive cars? Crashes are precipitated by flawed driving. But if we freeze the frame right there, there is no crash. Something else comes into play – the non-deterministic split-second layout of other vehicles and objects within a critical range of the errant vehicle, plus the instantaneous reactions or non-reactions of other drivers. Let's call this the critical core reaction.

Much more deterministic vehicle and infrastructure factors then come into play to complete the crash event and influence the amount of harm caused. But when does precipitation become causality? In every case there is a vivid and unusual story to be told, and the all-important critical core is more random than deterministic. This makes reverse engineering and prevention difficult. We really don't know about the crashes that almost happen – we only know a little about the ones that do happen. In the case of a vehicle driver failing to respect intersection controls, the precipitating flaw has been placed on the table. But Nader's two crashes will only result if there is another critically placed vehicle proceeding across the intersection, and no evasive action is taken.

2.5.3 Moral Responsibility, Causal Responsibility and Moral Luck

In February 2017, an aging Oxford academic ran a red light and struck a 9-year-old boy who was thrown under the car traveling at 20 mph but was miraculously unharmed (11). The professor was found guilty of dangerous driving and the police determined that he was oblivious to the crash and continued driving – because his eyesight was basically non-existent. Enter major irony: the guilty party had a long career as an expert in causation and moral responsibility in the law. His peer-reviewed writings included "it is a civil wrong to cause injury to another while driving a vehicle". (Nevertheless, he had pleaded not guilty.) He had also written that "many actions are regarded as wrongful whether or not they cause tangible harm".

This case touched on several branches of ethical and moral thought. The consequentialists would reduce the weight of the professor's actions because the boy was not harmed. The deontologists would place additional ethical weight on his neglectful approach to his vision – which would be a factor each and every time he drove – as well as his obvious failure to obey traffic law. The libertarians would rejoice in the fact that he still enjoyed automobility at the advanced age of 97 years. But they would not like the subsequent invasion of his privacy in the form of an article in the *Oxford Mail* – this could have been the greatest punishment of all. Philosophical principles are rarely applied in road safety literature and practice, as we hew closely to the matter of fact approach of police investigations. We need to think a lot more about the ethics of crash causation and responsibility for crashes.

Note that this eminent philosopher, who had been published in *The Stanford Encyclopedia of Philosophy* (12), had chosen to use the terms "wrong" and "wrongful". This invokes the application of virtue ethics in driving and is the complete philosophical opposite of utilitarianism. At the very least, road safety's extensive reliance on the latter school of thought is incomplete or even neglectful.

According to *The Stanford Encyclopedia of Philosophy* (12), the notion of moral luck arises because we are morally assessable only to the extent that the subject under assessment is within our control. This principle surely applies to highway crash causation. We may also make the distinction between moral responsibility and causal responsibility. Moral responsibility has the specific meaning of a person having a suitable relationship to the matter under consideration – that they have the right social footprint, capabilities and involvement in the incident to be under the spotlight. On the other hand, someone bears causal responsibility if they are the salient reason for an occurrence. This is a much more clear-cut assessment, more specific in scope. In the case of our Oxford scholar, he is morally responsible for sure – probably more so than many of his fellow Oxfordians, who lack his calibrated insights in matters of ethics. Many would also assess him as causally responsible, but for what exactly? Certainly a breach of traffic law, but also an alarming near-miss. He is definitely an icon of moral luck.

When the head of General Motors gave his verdict on motor vehicle crash causation in 1967, he probably meant that drivers are *morally* responsible for precipitating all crashes. They passed a driving test and acquired a motor vehicle with which they probably developed a reasonable level of familiarity and competence. They are aware of the potentially serious consequences of crashes for themselves and their family. So they could be expected to refrain from expressing personal flaws when they're out driving on the public roadway. But the consequences of a crash have little to do with its precipitation. Nader had actually raised the issue of responsibility in causing the 1960s onslaught of serious and fatal injuries, and GM had deflected the issue to crash causation. All drivers do have a moral responsibility not to precipitate crashes. However, once we enter the always-unique critical core, the driver cannot be held solely, or even largely responsible. Our main protagonist, and other drivers in that core, cannot be expected to execute effective emergency evasive action. Many deterministic factors will then come into play in the actual crash event, but we would not think of the vehicle or roadway as being causally responsible.

2.5.4 Reverse Engineering Complex Crashes

NHTSA's early-century naturalistic studies, and those of Transportation Research Board's (TRB) SHRP2 study, succeeded in recording a decent sample of actual crashes. Those studies excelled in showing all of the observable events taking place inside and outside the vehicle in the immediate lead-up to the crash. But in terms of each crash being a unique narrative, often with a long prior timeline, naturalistic studies still have their limitations. They cannot show crashes as life events that happen to involve driving roadways in cars. So NHTSA carried out a large in-depth study called the National Motor Vehicle Crash Causation Survey (13), presented to Congress in 2008. In-depth studies are normally small in terms of sample size because a multi-disciplinary research team needs to attend the crash site in a very timely manner and interview participants, witnesses and police, and then a lot of legwork is needed to fill in the entire crash narrative and backstory. This study was a milestone in that it investigated a massive 6,950 crashes over a three-year period. Among many data elements, NHTSA was searching for the critical pre-crash event, and the reason for that event.

NHTSA is definitely not given to loose talk or hyperbole. The agency had commenced its existence in 1967 and had developed hardened methods of bringing better and better vehicle safety rules to bear. But their descriptions of prime pre-crash reasons stay well away from the moral dimensions. Looming large among pre-crash events were driver failures in negotiating intersections and in staying in their driving lane. But NHTSA was kind enough to characterize critical reasons as "driver errors". They described these errors as belonging to one of the following classes: recognition, decision and performance. In reality, the list of errors thus compiled speaks convincingly to driving flaws that precipitate crashes. NHTSA saw these critical reasons as triggers for a new assault on crash avoidance, particularly through the expanding use of newly developing driver assistance technologies. But there is more to this than correcting driver errors.

We still have great difficulty in reverse engineering from a whole set of observable pre-crash situations to a limited number of effective applications for crash avoidance. It's a bit like turning a bucket of Chick-fil-A pieces back into a chicken. Too much processing and homogenization of the original crash has occurred, and the pieces packaged with too many other crashes. For the past 60 years, we have relied on engineering, technocratic and bureaucratic descriptions of crashes and crash causation. It is time to apply a more philosophical lens to the problem. In doing so, we need to get beyond the road safety bubble and think about safety as an attribute of the transportation system. Crashes are caused at the intersection of the transportation system – driver, vehicle, roadway and environment – and the powerful life narratives of drivers and occupants. Crashes are caused by driving flaws, precipitating factors and contributory factors. Some of these factors are so numerous, interactive and obscure as to become "accidental".

2.6   THE INCIDENTAL NATURE OF DRIVING

According to the U.S. Bureau of Labor Statistics (14), daily travel by adult citizens is not considered to be a primary task. It is considered to be a subset of other major activity categories such as working, purchasing, education, sports and caring for people who are not a member of the immediate household. Driving is largely an adjunct activity and clearly does not command the energy and attention of many people. There is little sense of achievement or personal pride in driving per se. With the possible exception of teen drivers' parents, citizens do not consider driving to be risky. As a life-long activity that consumes so much money and time and is mostly a means to some other end, driving is unique. It is a life choice made by a majority of the populace, but the risk is not zero.

2.6.1 INFREQUENT CONSIDERATION AND ACTIONS RELATED TO SAFETY

Many people take safety into account when they purchase a new vehicle. This may be as simple as buying a bigger, heavier SUV that could protect the family better in a crash. And the SUV may have the bonus of a higher seating position, and superior field of view. NCAP safety ratings may come into the decision; these are based mainly on crash-worthiness but also include crash avoidance features. The vehicle may contain advanced driver assistance system (ADAS) technologies, but there is no agreement on a minimum set. For some, the age of the vehicle correlates with the extent and quality of safety features, such as more airbags. So the owner may think that a newer car is generally safer. Subsequent to the purchase, the owner may opt for usage-based insurance (UBI) where proprietary sensors gauge some pertinent aspects of the driver's style, including harsh braking and cornering. Then safer driving metrics result in lower premiums.

Although many people will develop the character trait of buying a safe vehicle, from the perspective of the occupants, few would consider the aggressiveness of their own vehicle toward other cars. There are no standard tests for a car's outward aggressiveness. Beyond an awareness that a full-size pickup may cause more damage to a mid-size or small car – through mismatch of bumper heights and sheer mass – there is no formal or semi-formal way of thinking about this. The thoughtful driver would be at an even greater loss to select a route composed of safer roads. Some may consciously decide to conduct a local trip in an urban area using lower-speed surface streets, rather than jumping on and off the freeway, but such safety values are too vague for most people.

The vehicle owner wishing to develop a habit of safe driving – beyond obeying the traffic law – has limited options at their disposal, and most involve matters of vehicle selection and technology add-ons, and occur well upstream of any specific trip. Many people who feel that they possess virtuous attitudes to road safety may go out of their way to be considerate to other drivers. It is unlikely that they never violate the traffic law, but they probably feel that their violations are less risky and frequent than those of other road users they interact with. Over many decades we have all come to expect continuous safety improvement on the part of the automotive industry. So we may feel that we have limited ability to continuously "work the virtues" of safety – it is more a case of "set and forget" until I buy my next car.

2.6.2 MISSING PRINCIPLES FOR DRIVING ON THE COMMONS

Car drivers cause serious crashes through being uncaring and even negligent. It is remarkable that this is the case even under extenuating circumstances: even when surrounded by large, heavy, sharp-edged and unyielding trucks. And, at the other end of the scale, when weaving between people invading the roadway on foot – pedestrians.

Drivers are generally oblivious to some important principles that should apply to driving on the commons, but what are those principles (aside from the prevailing traffic law)? In her book *Technology and the Virtues* (15), Shannon Vallor proposes a set of "technomoral" virtues that should be cultivated. Her main focus is online morality, but some of her modern attributes of approbation include self-control, or being the author of our desires, empathy and civility, or making common cause. While high moral principles are rarely applied to driving, it does seem that we are missing the gene for an active driving ethic. Retreating all the way back to Aristotle, we are reminded of several practical personal traits of value, including phronesis (practical wisdom) and techne (craftsmanship). What has gone wrong? Why don't we drive better?

Western societies' tried-and-true ethical principles are all about actions: are they right, are they good? Rather than waiting to see whether we get home alive, we need to assess our travel choices before we go. But we don't want to have to carry out such an assessment for every trip. So we need rules of thumb. Given the life-long aspect of driving, these rules may be consciously formulated well upstream or may arise unconsciously over a period of time. And it may be easier to formulate rules for others – such as teens being banned from taking the most powerful vehicle in the garage – than for ourselves.

Are Shannon Vallor's technomorals at all applicable to driving? Does the morality of a person's activity online bear some similarities to that of their approach to ground transportation? They both represent a commons where people have a wide scope, and perhaps an incentive, to behave selfishly. Some would regard the physical commons and the online commons as two sides of the same coin. For example, activities accomplished online may reduce the need for personal travel. The new normal, post-pandemic, of working from home for at least part of the week is a great example of this symbiotic relationship, even though public transit is affected more than driving. People are now able to satisfy their need for access – to employment, entertainment, goods, services and many others – using either, or both, of these commons. What these modalities have in common is that they represent a derived demand, not a primary demand. This probably has a big effect on people's ethical mind-set. Most people are not professional drivers, nor are they professional bloggers for that matter. So we do not bring a high-minded approach to either modality.

The modern virtue of empathy is a case in point. Online technomorals need to deal with emotional issues running the whole gamut to psychological damage. Driving is much less threatened by mental abuse and manipulation applied by one person or group to another, although road rage does occur. However, road rage is a passing encounter compared with the locked-in venality that can be inescapable online. Beyond these important differences, both forms of commons benefit greatly from the personality traits of self-control and civility. In Aristotle's formulation of virtue ethics, part of the bedrock of Western society, people may aspire to practice and eventually acquire traits of character that embody specific moral and intellectual virtues. However, unlike our modern utilitarian instinct to maximize, Aristotle's virtues seek the Golden Mean and avoid the vices of excess and deficiency. So in the sphere of self-control, we may seek the mean of being assertive and avoid the excess of being aggressive, on the one hand, and the deficiency of passivity on the other. A person displaying the virtue of self-control would certainly not fall asleep or allow themselves to become distracted. There is no doubt that our tried-and-true utilitarian outcome of fatal crashes quantified would benefit greatly from drivers being programmed with a good and civic will. Or by vaccinating drivers against their known failures by fitting their vehicles with a well-informed selection of crash avoidance systems.

Our society needs to understand that a determination to make common cause with other road users would be an indicator of a virtuous person behind the wheel. But how exactly would that determination by employed or recognized? The virtuous person who invests in a well-informed selection of crash avoidance systems is less likely to collide with another vehicle. But that virtuous model needs to act as an inspiration to others, so that more and more vehicles are less likely to crash, and the traffic system as a whole is a lot less likely to host collisions. We have arrived at an important ethical distinction. The necessary collision avoidance systems already exist. But in the old regulatory world, we would need to make certain systems mandatory. This involves great challenges in economic justification of complex software that needs continuous improvement. This has not happened and may not happen. In a new technomoral world, the required technologies would be brought by vehicle owners who demand to be in control of their driving. And the virtues of self-control, driving wisdom, moderation and civility would be appreciated by one's peers.

2.6.3 HARMFUL MIND-SET OF DRIVING

The sheer extent of obliviousness shown by drivers on a wide scale, and consistently over many years, reveals a harmful mind-set. This mind-set has now been shown to precipitate serious collisions. It also escalates to negligent and aggressive driving behavior that bears causal responsibility for many fatal crashes. Much of this damage to the fabric of our road transportation system has occurred within the letter of the traffic law and right under the nose of the road safety enterprise. To many, driving has become a daily chore that is not worth caring about.

2.7 TRAFFIC AS A CAUSAL FACTOR

In the traditional sense of crashes between motor vehicles, injuries and fatalities are all about drivers, and sometimes also about their passengers – all of the humans in the vehicle. Humans in the vehicle impact the interior of the vehicle when it is involved in a collision; in the early days of zero restraint, they sometimes impacted their way out of the vehicle and then suffered collisions with fixed objects. Each time the human experiences such impacts; they are attacked by concentrated forces that cause injury when the forces exceed a certain threshold. This has been described as a vehicle collision that is followed by a human collision. Just as it has proven possible to reduce the severity of the human collision, it is also possible to mitigate the vehicle collision.

2.7.1 FOCUS ON ONE DRIVER AND ONE VEHICLE

The mid-century epidemiologists focused on the human collision – commonly called occupant protection. Subsequently, safety researchers also focused on the vehicle collision, from the perspective of complete avoidance – commonly called "crash avoidance". Most observers could see that the latter was going to be much more difficult, fearing the insurmountable obstacle of driver behavior. Some thought crash avoidance was impossible. But it still makes sense to seek measures to reduce the severity of the vehicle collision, even if it can't be totally avoided. In a chain of violent events, this is going to help the vehicle occupants.

The simple logic of occupant protection involved advantageous location of the human within the vehicle, occupant restraint, interior strength, compliance of interior surfaces and structural management of the crash forces acting on the vehicle. It was a given that the occupant would always remain inside the vehicle, in their assigned seat. The seat would also remain firmly attached to the vehicle structure. The correctness of all these rules of thumb was verified in crashlab tests, where human decelerations – the very agents that tear at organs, flesh and bones – were measured and forced into engineering logic.

For decade after decade, the eyes of the road safety fraternity were firmly fixed on life-saving responsibilities of the vehicle and the driver. What else was there? What other parts are needed to make up the road system? In an abstract way, safety researchers would add "the environment" or "the infrastructure". Strangely, the other candidate – the traffic – wasn't seen as a direct threat to the driver. And traffic, in the sense of helping or hindering mobility, is often considered to be an outcome of the system rather than a contributing component, an input. Obviously Vision Zero has tried to change that by noting that road and traffic organizations have been able to avoid serious scrutiny up to now – and now need to be held responsible. In fact, this was pointed out more than 20 years ago – that responsibility for road safety needed to be fully shared. But Vision Zero has failed to provide any major breakthrough in countermeasures rooted in the trafficway.

Extensive collection and analysis of crash and injury data has underpinned scientifically based "two collision" countermeasures. That refers to the first "vehicle" collision and the second collision experienced by the driver. We tend not to be fully specific about the "other vehicle" in the vehicle collision because it may be a roadside object in the case of a "single-vehicle crash". It may even be the surface of the roadway in the case of a single-vehicle rollover collision. But surely Vision Zero thinking would require a serious search for the specifics of the first collision, comprising the subject vehicle (Vehicle A) and the "other" vehicle (Vehicle B). Such knowledge would surely involve dynamics of the trafficway – the combined influence of the fixed infrastructure and other vehicles moving within the traffic – and would result in new countermeasures.

2.7.2 CRASH TYPES FROM THE PERSPECTIVE OF THE TRAFFICWAY

Perhaps some help is at hand. NHTSA carried out a comprehensive study, released in 2007, entitled *Pre-Crash Scenario Typology for Crash Avoidance Research* (16). This was "researchers' research". It was intended to prioritize crash issues and scenarios for attention in the search for all-or-nothing crash avoidance measures. However, it also casts some light into the shadowy corners of the trafficway, as it affects the first collision. Rather than reporting numbers and frequencies of fatalities, NHTSA used the more nuanced metric of "functional years lost", which is dominated by fatalities but includes all levels of injury and disability.

The trafficway scenarios most injurious to humans in two-vehicle collisions were expressed in the following crash types supplemented with "over-represented" circumstances and behaviors:

- Opposite direction without prior vehicle maneuver (11.6%)
  - Rural 55 mph speed zone
  - Two-lane undivided roadway
  - Drifting into oncoming lane
  - Either straight or curved
- Straight crossing paths at non-signalized intersections (11.6%)
  - Urban or rural 25 mph speed zone
  - Stop sign
  - Stopped before proceeding in error
  - Side impact from cross-wise vehicle
- Lead vehicle stopped (10.9%)
  - 35/45 mph speed zone
  - Signalized intersection
  - Or intersection with no controls in either direction.

By comparison, the order of these scenarios by frequency of occurrence is reversed, namely:

- Lead vehicle stopped (20.5%)
- Straight crossing paths at non-signalized intersections (6.5%)
- Opposite direction without prior vehicle maneuver (2.4%).

So the scenario of a following vehicle impacting the rear of a stopped vehicle is extremely common and is also one of the most frequent sources of highly injurious "second collisions" – the human impacting the interior of the vehicle.

NHTSA also provides useful results for multi-vehicle collisions, defined as more than two vehicles involved. With respect to human injury, the order of crash scenarios was as follows:

- Lead vehicle stopped (29.6%)
  - More common at signalized intersections
- Lead vehicle decelerating (13.3%)
- Opposite direction without prior vehicle maneuver (11.7%).

It is noticeable that, with more vehicles involved, the series of collision is more related to the need to decelerate – caused by a stopped or slow-moving vehicle. In terms of trafficway dynamics, the scenario of a stopped vehicle at a signalized intersection is by far the most damaging to humans in multi-vehicle crashes. This shows that traffic signals are deficient in effectively stopping more than two vehicles in the lane approaching the intersection.

The NHTSA data also paints a picture of just how tangled a web the trafficway is. First of all, the vast majority of light vehicle crashes (65%) involve two vehicles, while 28% involve a single vehicle, and 7% multiple vehicles.

This suggests a significant involvement of the traffic stream – that only a minority could be addressed by thinking about one vehicle and one driver only. It is also noteworthy that the number of injured humans per crash increased a little from a single vehicle to a two-vehicle event (from 0.38 to 0.44), but quite dramatically from a two-vehicle to a multi-vehicle event (0.44–0.97). In terms of overall functional years lost, the breakdown of human harm is as follows: 50.2% in two-vehicle crashes, 39% in single-vehicle crashes and 10.4% in multi-vehicle crashes. With 60.6% of human injury being caused in crashes involving more than one vehicle, one would be hard pressed not to conclude that traffic is a very important influence in road safety.

2.7.3   INFLUENCE OF TRAFFIC ON CRASH CAUSATION

How should we think about traffic as a factor contributing to the causation of motor vehicle collisions? The issue of traffic congestion, and delay in traffic, is surveyed each and every year by the Texas Transportation Institute (TTI) who report nationwide metrics including billions of miles traveled and billions of hours of delay caused by congestion. Much of the delay speaks to additional travel time when partly clogged traffic streams are forced to travel at speeds below the "free-flow" speed. Traffic does not need to be stationary to cause a huge element of delay, according to these types of measures. In 2019, before the pandemic cratered driving for a year or two, TTI reported a national total of 1,600 billion miles traveled and 8.7 billion hours of travel delay (17). So a trip of 100 miles that might have taken less than 2 hours at free speeds would take an extra half-hour under conditions of traffic delay. This means that a lot of driving is done under conditions where the single vehicle with driver is not "free" but is encumbered by other vehicles and drivers.

The NHTSA report separated out single-vehicle crashes (28% of all car crashes), two-vehicle crashes (65%) and multiple-vehicle crashes (7%). The high proportion involving more than a single vehicle (a massive 72%) surely demonstrates the importance of traffic in road safety, but in ways that are rarely expressed.

We should note that the only mention of crashes in the TTI mobility report is the critical need to tow away crashed vehicles that block lanes and create additional delay. But we may be on the cusp of a new era where most driving is encumbered by other traffic. We have already moved past the point of being able to solve safety issues by studying single-vehicle crashes. One important question arises: are single-vehicle crashes, two-vehicle crashes and multiple vehicle crashes just the same crash with extenuating circumstances?

If we go back to the seminal NHTSA report, the trafficway scenarios most injurious to humans in single-vehicle crashes were found to be as follows:

- Control loss without prior vehicle action (38.4%)
  - Rural 55 mph speed zone
  - Ran off roadway
  - Speeding
  - Straight roadway
- Road edge departure without prior vehicle maneuver (24.7%)
  - Rural, all speed zones
  - Non-junction
  - Going straight
  - (No recording of roadside objects)
- Pedestrian crash without prior vehicle maneuver (12.6%)
  - 25/35 mph speed zones
  - Urban
  - Non-junction
  - Going straight.

Comparing this pattern of critical single-vehicle crashes with the previously mentioned results NHTSA obtained for two-vehicle crashes, we see a totally different set of scenarios from the trafficway point of view. For single-vehicle crashes there are two simple rules of thumb:

- Stay on the roadway; and
- Don't hit pedestrians.

Both of these rules apply to drivers and to the driver assistance technology in vehicles. The road and traffic designers should definitely start to contribute with countermeasure development within their own purview. This would be a new role for these public sector players.

For crashes involving two vehicles or more, the cardinal rule of thumb for drivers and driver assistance technology is as follows: stay in your lane. But there are also equally important rules of thumb for road and traffic designers:

- Remove all stop signs with uncontrolled cross-traffic – we could describe this as asymmetric intersection control; and
- Make sure platoons of vehicles all decelerate and stop together at red signals (not just the lead vehicle).

It should be noted that the NHTSA report only scratches the surface with regard to the influence of the trafficway on road safety. Many of the NHTSA scenarios are lacking information because the biggest coding in many instances is "unknown". However, there is sufficient information to conclude that

- Collisions involving two or more vehicles are fundamentally different from single-vehicle crashes;
- Countermeasures to two-vehicle crashes need to take in the whole system: driver, vehicle technology and the trafficway;

- There is insufficient understanding of the system of collision because the necessary system data suffers from the myopia of post-crash evidence – as per police trying to attach blame to Vehicle A; and
- When we talk about "crash avoidance technology" we are conflating several scenarios: (1) single-vehicle-only crashes – a small part of the total road safety problem, and (2) Vehicle A in a multi-vehicle crash (even though Vehicle B may need different functionality from Vehicle A in order to also help avoid the crash).

2.7.4 THE DRIVING SYSTEM AFFECTING ROAD TRANSPORTATION SAFETY

In the past 20 years or so (2000–2019), licensed drivers and vehicle miles of travel (VMT) have increased by about 20%, while vehicle registrations have reduced somewhat. During that time, total fatalities decreased by about 12%. This occurred on a total road system that was relatively static in terms of lane-miles, which increased by about 7%. Clearly the latter is always a slow-moving measure. The average delay for a 100 mile trip increased from 26 minutes to 33 minutes – a traffic encumbrance penalty of 27% (17). The question arises: are we looking at a future where VMT increases inexorably while lane-miles plateau? Would that mean more dense traffic streams and more multi-vehicle collisions? This would be an adverse safety trend because multi-vehicle collisions cause more human injury per collision, and it is harder to counter their occurrence.

We will not know the answers to these questions while we know so little about the "system" of multi-vehicle collisions and the influence of trafficway conditions. The only real-time data we have as the crash unfolds is from naturalistic data with a single-vehicle/single driver focus, and even then these are low-severity crashes, studied in relatively small numbers.

While developments in ADASs will continue to address single-vehicle and "Vehicle A" crashes and will need to be hardened so that the vehicle reacts automatically in certain scenarios, road and traffic designers need to work on totally new ways to counter multi-vehicle collisions. A full-court press is needed for smarter signalized intersections. It will not be sufficient to wait for ever-smarter ADAS systems that are able to compensate for deficiencies in traffic control at intersections.

The influence of traffic in causing harm in collisions is a green field of safety waiting to be explored. We are so wedded to Vehicle A and its driver that this vehicle is assumed to be the cause of the collision. We know very little about the role of other vehicles in the vicinity. How does a vehicle get to be "Vehicle A"? It may be the only vehicle with seriously injured or killed occupants. It may appear obvious to police that this vehicle was in the wrong: for example, on the wrong side of the road or running a red light. Thus, exterior harm plays an important role in tracking down crash causation. But if we look closely at NHTSA's trafficway scenarios, interior harm – lapses in drivers' mental processes – plays a huge part. Drivers frequently stray to the wrong side of the road, fail to stop when the preceding vehicle is stationary and fail to check cross-traffic at uncontrolled intersections. In a certain sense, Nader's two crashes are really three:

1. Mental "crash" on the part of the driver;
2. First physical crash; vehicle to vehicle or vehicle to object; and
3. Second physical crash; driver with vehicle interior.

The mental crash may be deliberate, irresponsible or negligent. Nevertheless, the driver causes a line to be crossed. They enter a conflicted state; rationality and competence are damaged, and recovery is destroyed.

Such considerations should become central in the pursuit of the "system responsibility" advocated in Vision Zero. But current data will probably not support this approach.

2.8   ARE LARGE AND HEAVY FREIGHT TRUCKS AN EXCESSIVE THREAT?

With very few exceptions throughout the national road network, cars share the roadway on a daily basis with large and heavy freight trucks. Apart from curfews in some big cities, freight trucks enjoy the same unfettered access to roadways as personal vehicles – in all parts of the country. But right off the bat freight vehicles were treated differently by regulators, not just because they are so much larger, heavier and stronger, but because they are using the commons for direct commercial gain. Many mid-century economists cut their teeth on truck regulation in relation to market access. Many states instituted commodity-specific regulations that excluded over-the-road trucking from carrying certain freight. This was part of a tendency for over-the-road carriers to haul higher-value products, often of lighter density. From that time, the sector of freight operations has been fair game for regulators, and subject to an expanding array of safety regulations.

2.8.1   HEAVY VEHICLE ACCESS TO THE ROAD NETWORK

The network access of large and heavy trucks is largely unrestricted up to certain weight and dimensions, typically 80,000 lb of gross weight and a box up to 57 ft in length. Truck size and weight have long been limited under federal and state regulations because open-ended misuse of trucks would place economic advantage above safety and the wear and tear they cause for road pavements and structures like bridges. Trucks and their drivers, and the companies that own and operate them, are subject to many safety rules. Drivers need special licensing and are subject to limits on the hours they spend behind the wheel. Heavy truck speed limits lower than car limits apply in most places. As well as being subject to vehicle safety standards administered by NHTSA, heavy vehicles are subject to roadside inspection, especially of brakes and the like. All states operate a network of truck weigh scales, and drivers' records of recent working, sleeping and recreational hours are checked. Some own-account haulers and carriers choose to install technologies that monitor driver compliance with rules as well as performance of the driving task, adherence to traffic rules etc.

In 1999, Congress passed the Motor Carrier Safety Improvement Act. Commencing on January 1, 2000, the FMCSA came into being. This new federal agency was installed within the U.S. DOT, alongside NHTSA and other modal administrations. The centerpiece of its efforts to regulate the safety of carriers is the Compliance, Safety, Accountability (CSA) program that administers a scoring system for carriers. This is used to prioritize the lower-rated carriers for safety inspections and sanctions. In addition to such industry-specific rules, heavy vehicle safety has been researched in a manner similar to that pioneered by NHTSA for personal vehicles – cars, SUVs, pickup trucks etc. A number of crash studies have been carried out where the heavy vehicle is front and center as Vehicle A. Crash and injury rates have been determined and tracked for various weight classes of heavy vehicle. This sort of information would obviously be used to compare safety metrics for light vehicles and heavy vehicles. It would be used by state and federal governments, and many others, to maintain a certain degree of safety equity between those driving the roads for personal reasons, and those who drive aggressive-looking vehicles for a living.

2.8.2 RESPONSIBILITY FOR CRASHES BETWEEN CARS AND HEAVY VEHICLES

It may well appear that large freight trucks pose an unreasonable threat to other motorists. Deep-seated factors behind these thoughts might include unrealistic schedules imposed by carriers, tired drivers, aggressive driving, excessive speed, mechanical defects, outdated technology and inexperienced drivers. However, the University of Michigan Transportation Research Institute (UMTRI) worked diligently for decades to compile the facts. This included a full-time team of up to 20 phone interviewers who pieced together the backstories of many thousands of fatal truck crashes. As with all road safety professionals, the UMTRI team was very particular about terminology when it came to assigning blame. Among a multitude of factors that plausibly contributed to these crashes, the researchers looked for critical reasons, and whether this idea of criticality should be assigned to the truck driver, the car driver or both. However, the issue of responsibility needed to be resolved. The UMTRI researchers found that, in a very large majority of two-thirds to three-quarters of cases, crash responsibility was assigned to the car driver (18). The truck driver was only responsible in a minority of cases. This surprised a lot of people and still surprises people.

Annual U.S. deaths in large truck crashes equate to approximately 10% of all road fatalities. While this figure may not stand out as being excessive, certain aspects of car-into-truck crashes do cause concern. Even though the car driver is usually responsible, the person killed is almost always in the car. The sheer weight, strength and frontal aggressivity of large trucks are overwhelmingly damaging to cars in frontal or side collisions. When a car crashes into the side or rear of a semitrailer, the tendency for the car to run under the structure of the trailer causes horrendous, even macabre, injuries to occupants of the car. This causes trauma for all involved in the crash and its aftermath, including emergency and medical personnel, as well as police. The carrier also experiences significant negative effects on their drivers, operational personnel and management. There is an uneasy tolerance of high-severity crashes involving large trucks, along with disquiet about the fact that such vehicles ply the highways in a relatively unrestricted way, covering vast distances that may expand the exposure of personal vehicles to these types of violent crashes.

2.8.3 ECONOMY VERSUS SAFETY?

There is a general aversion to trading off human life versus economic gain on the nation's highway system, and this issue is probably closest to the surface in car–truck crashes. Even if the truck's role may often reside in just being there, ethics do arise in terms of unproductive "deadhead" miles and the role of just-in-time delivery, where companies prefer to have the freight in motion rather than stockpiled in warehouses. Such practices increase exposure on the roads, perhaps for questionable reasons. Another little-discussed moral issue arises when the truck driver avoids carrying out an evasive maneuver because he or she has an innate fear of leaving the roadway and rolling the truck. If that happened and they survived – by no means guaranteed – they may also anticipate a credibility issue with carrier management who say "and exactly which car did you avoid"? The ghost of Vehicle B again raises its head in preventing us from getting to the bottom of crash causation.

It would be wrong to conclude that large, greedy corporations are quietly adopting operational practices that broadly increase crash risks. It is well to remember that most of the large trucks on a given highway, on a given day, do not belong to large or even medium-sized corporations. They are owned by small, often family, operations who work extremely hard, competitive, long hours for a meager financial return. This explains why even those crash avoidance technologies that do work when placed in the cab of a large truck have been slow to penetrate the national truck fleet. There are good technologies, such as lane departure warning, that reduce crash risks for large trucks. The question is: how well equipped to figure out the payback period is an over-the-road owner-driver? This is an important issue for owner-drivers and small family fleets. While one partner is out driving, the other is managing the books while maintaining a level of concern about safety.

The awkward question of truck safety versus economics is also complicated by the fact that large truck safety is sensitive to the geometric standards of the roadway – much more so than cars. Large trucks are much more likely to be struck head-on by a car on undivided roads than on multi-lane, divided, limited-access facilities. Because the agencies that build and operate roads are part of government, they are not compelled to adopt particular standards within a certain time frame, and their practices may be hard to criticize from the outside. When the author and his Australian colleagues had occasion to compare the crash rates of large trucks in the United States against those in Australia, the much higher rate in Australia was correlated with the lower percentage of trucking carried out on motorways (19). If you move the nation's freight on lower standard roads, violent head-on car-into-truck crashes will persist and even increase in number whenever economic times improve.

2.8.4 HEAVY TRUCK VERSUS CAR CRASH RATES

From the standpoint of utilitarian ethics – by far the most common invoked in the practice of road safety – does it make sense to compare crash and fatality rates between the class of motor cars and the class of heavy trucks? According to the NSC each heavy truck averages about twice the mileage of a car, and heavy trucks are involved in a correspondingly larger number of fatal crashes (20). To the safety professional, this does not look untoward. However, using data from 2021, 72% of persons killed were occupants of the other vehicle, almost always a car. It is also notable that 74% of these fatal crashes occurred off the interstates. That is, on arterials and local streets.

The FMCSA's 2006 analysis of critical reasons for truck–car crashes – covering a range of severities, including fatal crashes – found that the car was assigned the critical reason in 56% of fatal crashes, and the heavy truck in 44% of such cases (21). The critical reasons for these crashes, whether they apply to the car or the truck, are similar. They cluster around non-performance (probably asleep), non-recognition of a risk (often distraction), bad decision (driving too fast) and poor performance (excessive steering or braking). For truck drivers, non-recognition and bad decision-making were the most common of these. This was also true for the car drivers. The only big differences occur in the smaller categories: car drivers virtually own falling asleep at the wheel and excessive avoidance reactions. We begin to see a picture of heavy trucks being driven too fast on lesser roads, by distracted drivers. The cars they collide with are also driven too fast, by distracted drivers. But some of these drivers are also fatigued and fall asleep in motion. And when they finally realize the imminent risk, a significant number over-react and make the collision worse.

When it comes to comparing the rates of collisions between two cars, on the one hand, and collisions between a car and a heavy truck, the two categories are remarkably similar. The road safety utilitarians' crashes per mile calculations don't need to treat heavy truck collisions as a separate category. Turning to what we know about the good intentions – or otherwise – of the two classes of driver, we see a similar level of negligence shown to the driving task. Neither class could lay claim to safe driving as a virtue. The question is: does the constant presence of heavy trucks represent a moral threat to the vast majority of drivers whose cars swarm all over our roadways?

2.8.5 MORALITY OF ASYMMETRIC INJURIES

First of all, we know that the worst form of harm – loss of life – is heavily biased toward the drivers of cars. Is it therefore morally wrong for truck drivers to drive too fast, pay too little attention or refrain from sudden avoidance? From a Kantian perspective, could we say that the actions of truck drivers always display a good will? We can't go back to our utilitarian selves to answer this question. We should consider the motives of the drivers and whether they act from duty. Truck drivers undergo special licensing requirements and are tracked when it comes to criminal convictions and traffic violations. They are monitored for aspects of their health that affect driving. Their hours of service are subject to scrutiny. And many receive training specific to the nature of the carrier's business and operational requirements and are subject to specific company rules.

All of these things affect their ability to keep driving and remain productively employed. Most drivers don't work for large companies. When it comes to owner-drivers, they may feel suffocated by rules and requirements and need to have a strong instinct for survival. All would be acutely aware of the significant value of their truck, trailer and cargo, and the fact that they are an important link in someone's supply chain. Suffice it to say that truck drivers are strongly motivated to keep their heads above water. They could not be expected to act upon higher moral duties, but they are expected to follow a lot of rules appropriate to their employment. They frequently infringe on these rules, including exceeding the speed limit.

It is notable that critical reasons for heavy truck collisions often include acts of omission, rather than commission. Failure to recognize an emergent risk involving a car, perhaps because the truck driver is distracted, is not the result of a specific action. Similarly, traveling at a speed above the speed limit is generally not a bad act but is part of the driver's mind-set to get the job done. The most morally deficient omission on the part of truck drivers is evasive action. The FMCSA data shows that the car driver is 3.5 times more likely to over-correct than the truck driver. Considering that heavy trucks are much easier to over-correct than cars – because they are less stable – the evidence provides a hint that many truck drivers are not helping to avoid the collision. Inner thoughts about such decisions that amount to selective harm are unlikely to be expressed.

2.8.6   ETHICS OF INTERFERING WITH THE CRASH SEQUENCE

Philosophers have long considered the problem of *doing* harm versus *allowing* harm, and many argue that there is a crucial moral difference between initiating a sequence of events that kills someone and failing to prevent such a sequence. Philippa Foot, who originated the Trolley Problem in the 1970s, makes a distinction based on whether such prevention is done using positive rights or negative rights. Positive rights allow one to interfere in a sequence of events, while negative rights act against interference and tend to be stronger. If a heavy truck is proceeding in its correct lane on an undivided two-lane roadway, and a car is proceeding head-on toward the truck in its wrong lane, the car driver will be killed if the truck driver takes no action. Alternatively, the truck driver interferes by steering onto the shoulder of the roadway and the collision is avoided. But another harmful sequence is initiated if the driver loses control of the truck, perhaps leading to a rollover that destroys the cab and kills the truck driver. Or, there may be another vehicle pulled over in the shoulder, with which the truck collides and kills its driver. In addition to causing deaths, such additional sequences result in significant vehicle and cargo damage, disruption to livelihoods, supply chain interference etc. All of these events add to negative rights against interference.

Another philosophical framing of the crash considers whether the interference contributes to the success of the original scenario, and many more scenarios like it, or not. If we wish to have a reliable, highly repeatable, situation whereby heavy trucks and cars travel safely in opposite directions on two-lane undivided roadways, cars must stay in their lane. The truck driver's interference of steering onto the shoulder does not contribute to the desired scenario and is therefore unethical. Immanuel Kant insisted that all duties and obligations derive from the categorical imperative, the core concept of deontology. Kant believed in the morality of an individual's good acts being worthy of a universal law. Would "all vehicles on two-way, two-lane roads must stay in their lanes at all times" be such a law? Of course, this would prevent vehicles from overtaking one another on these types of roads. And we would need to be sure that there would be no need for exceptions – otherwise Kant would not accept it as a universal law with consequent moral standing. Road safety practitioners would certainly favor universal laws.

Another part of Kant's categorical imperative is treating all human beings as ends in themselves, not as means to an end. In the case of car drivers, they are behaving as part of a successful life as they see it. They are therefore willing to devote a large segment of their income to driving. They do not treat other car drivers as means to an end; in many different ways, they demonstrate a community of purpose interspersed with moments of great selfishness. But truck drivers are very different because they are at all times acting as means to other people's ends – hauling other people's goods. As such, they are not living their best lives while they are driving. The attitude of truck drivers to car drivers may reflect the disdain of the professional for the non-professional. In reverse, car drivers seem to resent being held up by big trucks and to shy away from close proximity to such large, heavy, aggressive-looking vehicles. Good will out on the road is therefore impaired for all truck drivers, and for all car drivers interacting with trucks – pretty much all drivers. It is in everyone's interest to find ways to bridge this divide.

2.8.7 Responsibilities of Car Drivers in the Presence of Trucks

What is to be made of some key failings of the drivers of cars involved in collisions with heavy trucks? These include falling asleep, failing to grasp a risky situation, driving too fast and over-reacting when they sense that a crash is imminent. Such failings certainly reflect poorly on their work ethic (as drivers), self-control and care for the well-being of others. And all of these character faults are reflective of a general sense of detachment of people when they drive. The philosophies of human beings as rational creatures motivated to act in certain ways – and those actions being fair game for moral scrutiny – do not fully explain these phenomena that apply, and prevail, prior to road crashes. There is no evidence that the presence of large trucks causes any change in the behavior of car drivers – no evidence of excessive caution or fear. Those of us who have observed hours and hours of face video showing people driving see little signs of life, emotion or application to a task. People's thoughts are elsewhere, and if not mulling some unrelated life experiences (positive or negative), making themselves available to be distracted.

Despite troubling aspects, the harm caused by the drivers of heavy freight trucks is not morally deficient relative to that caused by car drivers. Neither class of driver practices virtuous driving nor is particularly worthy of praise. The checks and balances introduced by organizations like FMCSA have succeeded in producing an equilibrium, and a sense of safety equity. As with road safety in general, the biggest source of safety improvement resides with personal drivers, starting with self-control and civility.

2.9   SEVERE CONSEQUENCES FOR VULNERABLE ROAD USERS (VRUs)

The other great asymmetry in road safety is that between motor vehicles and unprotected people who seek to share road space with cars, or whose daily purposes cause them to intersect with moving vehicles. So-called vulnerable road users (VRUs) include motorcyclists, cyclists, practitioners of micromobility (such as scooters) and pedestrians. While the road safety profession cut its teeth on motor vehicle collisions, there is increasing awareness of the fatalities and disabilities visited upon unprotected road users. In recent years, battery electrification has increased the popularity of higher speed "e-bikes" and new personal machines such as e-scooters.

VRU safety has gradually received more attention in the United States and other rich countries but has long been a major issue on a global scale. This is because the World Health Organisation states that "more than half of all road traffic deaths are among vulnerable road users: pedestrians, cyclists and motorcyclists". VRU safety takes on additional significance because vulnerability is often made more acute by factors of age and demographics. Children and the elderly are more at risk when using roads in an unprotected mode. And poorer citizens are more likely to be VRUs, more often, given their need to walk or bike to access transit networks.

2.9.1   RISING FATALITIES OF PEDESTRIANS

The early days of professional road safety, including the creation of NHTSA, focused on protecting vehicle occupants in collisions. NHTSA's statistics (22) show that the proportion of fatalities occurring inside the vehicle was 72% in 1980, and this figure had reduced to 66% by 2019. The proportion of fatalities outside the vehicle showed the opposite trend, increasing from 28% to 34% over the same period. This does not refer to vehicle occupants being thrown out of the vehicles, but to unprotected persons being struck by moving vehicles. It is notable that, in 2019, a full one third of all traffic fatalities occurred outside the vehicle and, by inference, were struck by a vehicle. After years of lesser numbers, VRU annual fatalities reached 12,000 through the period 2016–2019, the highest level since 1986.

When it comes to VRU fatality rates, NHTSA uses the metric "fatalities per 100,000 population". Obviously, more engineering-style metrics based on physical exposure are not available. It is concerning that the VRU fatality rate has been increasing since the turn of the century. For pedestrians, injury-reducing measures applied to the front ends of cars are reasonably well developed and a number of tests are in wide use along with desirable design principles. Several pedestrian impact tests are incorporated in the European NCAP safety ratings, for example. For the pedestrian, there is a multi-stage injury event, starting with leg impact, then upper body and head impacts with the hood and windshield, and finally ground impact after the vehicle has come to a halt. Most pedestrian fatalities are caused by head injuries and many disabilities are caused by first-impact leg injuries. Less is known about vehicle countermeasures specific to cyclists and motorcyclists.

From the perspective of the infrastructure – roads and streets – newer layouts attempt to separate VRUs from cars in both space and time. Certain principles have been established by the FHWA and the 21st century saw the advent of Complete Streets. Under this rubric, hundreds of jurisdictions across the country have implemented street, sidewalk and transit stop designs that are intended to safely accommodate all modes. These designs usually see some car lanes given over bicycle lanes.

Understanding the circumstances of pedestrian fatalities requires local knowledge of cities and urban areas. Mass crash data tends to be less specific for pedestrian fatalities than for those of vehicle occupants. It is therefore much more difficult to generalize about the events immediately preceding a pedestrian fatality than it is for a fatality involving motor vehicle occupants.

New York City sustains the largest number of pedestrian fatalities of all U.S. cities. In 2017, the City of New York released an analysis of pedestrian fatalities (23) that occurred during the three-year period spanning between 2012 and 2014. First of all, pedestrian fatalities represented more than half (56%) of all traffic fatalities. Road type was a primary variable in the analysis. This included highways, arterials and local roads. Highways allow only motor vehicles and speed limits are the highest. Arterials carry large volumes of traffic and are wide, with traffic signals. Local roads have the lowest speeds. More than half (58%) of pedestrian fatalities occurred on arterials and 24% on local roads. Almost half (47%) are caused by irregular crossing action. Namely, crossing against the light (midblock or intersection) or crossing away from an intersection. Striking vehicles are predominantly SUVs (31%) and cars (28%), probably reflecting the mix of personal vehicles on city streets.

2.9.2 MOTORCYCLE SAFETY

Motorcycle safety is a longer-standing issue in road safety and is probably the most difficult of the traditional streams of safety intervention. Motorcycles operate in and around general traffic across the full range of roadway environments, including the higher speed zones. In 2019, there were 5,044 motorcycle crash deaths, at the rate of 25.6 per 100 million miles of travel (24). This rate is an order of magnitude greater than that for cars. One of the principal causes of motorcycle collisions is that motorcycles are unlikely to be reliably detected by car drivers. Unfortunately, both the number and rate of motorcycle fatalities have been increasing since 2009. More than half involve a motorcycle and one other vehicle. The most contentious motorcycle safety issue of the 21st century has been the wearing of helmets, with many states repealing existing mandates. Nevertheless, almost 60% of these fatalities involve helmeted riders.

Motorcyclists provide a unique perspective on user attitudes. Personal freedom is a strongly held value, as evidenced by the roll-back of helmet laws in the United States in the early 21st century. And motorcycle interest groups in other countries have long advocated for safer traffic environments and more consideration exercised by car drivers. In Sweden, the home of Vision Zero, motorcyclists have proclaimed the need for a safe system – for motorcyclists. Motorcyclists tend to be extremely aware of the dangers around them and are often proud to display their riding skills relative to their semi-somnambulant counterparts behind the wheels of motor cars. Splitting of lanes on busy, high-speed LA freeways is a vivid example of their attentional fortitude. Motorcyclists like to demonstrate resilience and self-reliance and many are not amenable to safety advice or protective measures developed by governments.

The bundling of pedestrians, motorcyclists and cyclists as VRUs obscures the different safety solutions needed for each. They are not a single cohort when it comes to injury analytics and countermeasures. Crash rates based on exposure only make sense for motorcycles, but not pedestrians and cyclists. Specific vehicle countermeasures make sense for pedestrians, but not for motorcycles and cyclists. Improved behavior on the part of motorcycle riders could be part of a safety solution for motorcyclist-VRUs, but countermeasures for pedestrians and cyclists based on education and enforcement are less promising.

2.9.3   NEED FOR MORE AND BETTER CONSIDERATION OF VRUs

VRUs are resistant to the utilitarian philosophy generally applied to motor car crashes – that of controlling their consequences – because there are limited options for protecting them once a collision occurs. The limited effectiveness of motorcycle helmets is a case in point. A distinguishing feature of VRUs is that they are all relying on more and better consideration from the drivers of motor cars. This brings us over to the realm of civility and ethical intent, rather than the utilitarianism upon which much of road safety has been based. VRUs have an increasingly important place in road safety, but they don't fit the established modes of countermeasures. The frequent usage of the term "vulnerable" implies value judgments. We are obliged to make those judgments more explicit.

The ethical school of deontology, as presented by Immanuel Kant in the 18th century (25) and espoused by Bentham and Scruton, requires an action to be motivated by good will. While it would be hard to apply this test to every driver operating in the vicinity of VRUs, there are certain maxims that could refer to the often-exposed position of VRUs in the midst of traffic. One maxim states that each and every individual must be treated as an end in themselves, never as a means to an end. This may mitigate against street planning that herds pedestrians together for their safety but also makes things easier for motorists. Another relevant maxim is the "categorical imperative": it is permissible to cause harm in order to save more, but only if that harm avoidance part of the greater good. If a driver on a busy arterial is confronted with someone crossing the road illegally, it may not be ethical for them to stop suddenly and cause mayhem in the traffic stream. Even though compromises in urban environments may involve choosing between different forms of harm, they should always favor the cause of motor vehicles and VRUs co-existing safely. Deontology requires actions to be part of a larger moral duty.

If we put ourselves in the place of the VRU, we are hoping that oncoming drivers of vehicles are going to do the right thing. If they are of good character, they will exercise care and try to avoid causing any harm to the unprotected person in the roadway. This is entry level for a person of good character, but the great virtues such as wisdom, justice and courage are unlikely to enter their immediate calculations. So the VRU should be aware that the best they can expect is for the drivers to follow the rules. But those drivers' respect for the rules – possession of a driving license, ownership of a compliant vehicle, remaining un-distracted and alert, traveling at the speed limit, maintaining a safe headway and staying in the lane – may not be enough to save the irresponsible person in the middle of the arterial road.

We could certainly expect the designers and administrators of the infrastructure commons to have adopted clear design principles and strong rules for both drivers and pedestrians. They should also have proven success with certain values in their work: equity, fair dealing and truthfulness. These are all aspects of the headline virtue of justice, as is empathy. However, it does seem that the exercise of VRU safety falls short of being a moral calling: "doing the right thing" may not be enough of an answer.

2.9.4 NEED FOR WISE RULES

Rules are critically important to the reduction of harm to VRUs. Complex and varied scenarios should not be reduced to a numbers game. Let us consider one of the most common scenarios in pedestrian fatalities: a pedestrian crossing a busy arterial at a signalized intersection – but crossing on a red. The pedestrian is struck and killed by an SUV or car traveling through the intersection on a green signal. Assuming the driver was not intoxicated or speeding; the morality of their action would not be questioned. Even though the pedestrian is vulnerable and exposed, and the driver is well-protected in the vehicle, we struggle to be on the side of the pedestrian in this case.

In a less-common scenario, the pedestrian is crossing at a signalized intersection, and doing so under the auspices of a green light. The driver of the SUV or car is violating traffic law by proceeding through the intersection on a red light. This is clearly against the law of the land, and many would argue against the moral law. But not the highest moral law because it was not the driver's intention to kill the pedestrian. Unfortunately, the circumstances of these types of pedestrian fatalities are analyzed only to a minimal level, reflecting the inadequacy of data collection at the scene. Simple categories of road type, crash location and pedestrian action are insufficient to create meaningful "trolley problems". The pedestrian and driver are in highly asymmetric states when it comes to protection in collisions, but their intentions are almost impossible to deduce from the scene after the event. In the case of the vehicle failing to stop at the red signal and striking a pedestrian on a crossing, the vehicle is immediately assessed as Vehicle A and the police investigation ends there.

Of the mixed class of VRUs we have discussed, pedestrians are probably the most vulnerable to potential abuse by drivers of cars. However, there is more evidence of lapses on the part of pedestrians, than of the drivers of SUVs and cars. There are very few moral failures to be seen in the moment. It is in the nature of road crashes that small, sudden lapses by humans bring about horrendous injuries. And good will has little to do with prevention of the imminent crash. The safe co-existence of vehicles and pedestrians in cities requires wise rules conceived in a spirit of justice. Those rules should promote civility, empathy, honesty and self-control on the part of pedestrians and motorists alike.

2.10   SYSTEMATIC SOURCES OF HARM TO ROAD USERS

Bad driving is the greatest source of harm in our road transportation system. And the system invites more and more driving, so the original sin of careless driving is amplified more and more. The resulting harm, in terms of injuries and deaths, refuses to go away. Despite hard-won incremental reductions in driving risk the needle is not moving. Why has driving behavior become such a scourge? Distraction, seemingly innocuous in itself, has become a weakness present in the gene of driving and sets the scene for other, much worse, sins of driving. These include unresponsiveness, negligence, recklessness and aggression. While distraction potentially affects a majority of drivers, driving sins are limited to a smaller cohort. Such drivers cause harm to themselves and, to a lesser extent, other drivers and passengers. In the case of light vehicles – cars and SUVs – the harm is somewhat symmetrical for the guilty and the innocent.

In the moral sense, some of that harm is simply permitted to happen. For example, deficiencies in road and traffic design may turn distraction, or fatigue, into serious error. Rules may be unclear or irrelevant. The driver is morally responsible in that they should have known better. In other cases, fewer but still substantial in number, the harm is caused by the driver. For example, entering a crossing on a red and striking a pedestrian. Now the driver bears causal responsibility, whether they broke the traffic law or not.

In the case of asymmetric collisions – between a car and a heavy truck, or a car and a pedestrian – moral assessments could vary dramatically and be most severe when the more-protected partner bears causal responsibility. However, in the case of multi-vehicle collisions, we cannot be totally sure we have identified Vehicle A correctly and could be investigating the wrong driver. The time-honored methods of crash investigation are not conducive to ethical interpretation.

As we have observed several times, there is a great tendency to treat all aspects of road safety using the same utilitarian principles. It is assumed that continuous scrutiny of fatality frequencies will tell us how bad the crash problem is and will allow us to compare different aspects of the problem. But especially in the case of big city mobility, and the clear and present mix of vehicles and VRUs, we need to move to the "front end" of safety with system quality, interactive rules and a communal desire to do the right thing. In more and more cases, rules need to be crystal clear and based on morality. Road users are being harmed by the moral silence.

Vehicle-to-vehicle collisions are the greatest source of human injury and death in driving. The risks of such crashes have reduced over the modern era of road safety. These risks are seen to be so insignificant that the sheer quantum of driving has increased dramatically; there are few signs of societal discomfort with this status quo. As long as the risks of such harm – quantified against population and miles driven – are not seen to be increasing, other societal issues tend to assume greater importance.

However, the harm caused to VRUs, and especially pedestrians, when struck by motor vehicles has risen markedly. The mechanisms of these highly injurious events do not follow the established rules of utilitarian assessment and development of countermeasures. This trend undermines the credibility of the road safety enterprise and significant change is needed.

The scientific basis of road safety is also challenged by increasing traffic volumes and densities: most serious injuries and fatalities occur in traffic. Undue attention to "Vehicle A" and its driver in crash investigation is neglectful of the role of the traffic system in crash causation. It is no longer sufficient to study the behavior, control and integrity of a single vehicle and driver. Collisions between two vehicles cause significantly more harm than single-vehicle crashes, and there is evidence that three-vehicle crashes are more injurious again. Current traffic signals succeed in bringing the lead vehicle to a halt but are less effective for the following vehicle; this scenario looms large over all of road safety.

From our current vantage point, on the brink of new safety technologies and levels of knowledge, the main systemic influence on driving harm is the vast amount of driving we do. We do it in a largely off-hand way, and we don't do it for itself, but in support of our most important activities in life. At the outset of the modern era of road safety, economists warned that vehicle countermeasures would be offset by increased "driving intensity". This phenomenon may not have materialized in all of the ways that economists imagined but can be discerned in the relentless increase in miles driven. And this exposure problem is exacerbated by a consistent under-appreciation of driving risk: there is little evidence of driving moderation on account of safety.

REFERENCES

1. Nader, Ralph (1965) *Unsafe at Any Speed*. Grossman Publishers.
2. Drew, Elizabeth (1966) The Politics of Auto Safety. *The Atlantic Monthly*, October 1966 Issue.
3. Gladwell, Malcolm (2001) Wrong Turn. *The New Yorker*, June 11, 2001.
4. National Highway Traffic Safety Administration (2007) Pre-Crash Scenario Typology for Crash Avoidance Research. DOT-VNTSC-NHTSA-06-02.
5. Peltzman, Samuel (1975) The Effects of Automobile Safety Regulation. *Journal of Political Economy*, Vol. 83, No. 4. The University of Chicago Press.
6. Mill, John S. (1859) On Liberty. Cited in Macleod, Christopher, "John Stuart Mill", *The Stanford Encyclopedia of Philosophy* (Summer 2020 Edition), Edward N. Zalta (ed.). https://plato.stanford.edu/archives/sum2020/entries/mill/
7. Car and Driver (2016) Colonel John Stapp and the Fine Art of Crashing. December 3, 2016.
8. National Safety Council (NSC) (2023). Preventable Deaths. https://injuryfacts.nsc.org/. Accessed March 9, 2024.
9. National Highway Traffic Safety Administration (2023) The Economic and Societal Impact of Motor Vehicle Crashes, 2019. DOT HS 813 403.
10. Transportation Research Board (2014) Implementing the Results of the Second Strategic Highway Research Program. TRB Special Report 296.
11. Walker, W. (2018) Oxford Professor Who Ran a Red Light and Hit a Child Let Off with a Fine. *Oxford Mail*. April 23, 2018.

12. Nelkin, Dana K., Moral Luck. *The Stanford Encyclopedia of Philosophy* (Spring 2023 Edition), Edward N. Zalta & Uri Nodelman (eds.). https://plato.stanford.edu/archives/spr2023/entries/moral-luck/. Accessed on March 10, 2024.

13. National Highway Traffic Safety Administration (2008) National Motor Vehicle Crash Causation Survey. DOT HS 811 059.

14. U.S. Bureau of Labor Statistics (2022) Time Spent in Primary Activities and Percent of the Civilian Population Engaging in Each Activity, Averages Per Day by Sex, 2022 Annual Averages. www.bls.gov/news.release/atus.t01.htm. Accessed on March 10, 2024.

15. Vallor, Shannon (2016) *Technology and the Virtues: A Philosophical Guide to a Future Worth Wanting*. Oxford University Press.

16. Wassim Najm et al. National Highway Traffic Safety Administration (2007) Pre-Crash Scenario Typology for Crash Avoidance Research. DOT HS 810 767.

17. Texas Transportation Institute (TTI) (2019) 2019 Urban Mobility Report. The Texas A&M Transportation Institute & INRIX.

18. National Highway Traffic Safety Administration, Trucks in Fatal Accidents (TIFA) and Buses in Fatal Accidents (BIFA). www.nhtsa.gov/fatality-analysis-reporting-system-fars/trucks-fatal-accidents-tifa-and-buses-fatal-accidents-bifa . Accessed March 9, 2024.

19. Haworth, Narelle, Vulcan, Peter, & Sweatman, Peter (2002) Benchmarking Truck Safety in Australia. http://acrs.org.au/publications/conference-papers/database/. Accessed March 9, 2024.

20. National Safety Council (NSC), Road Users. https://injuryfacts.nsc.org/motor-vehicle/road-users/large-trucks/. Accessed March 10, 2024.

21. Federal Motor Carrier Safety Administration (FMCSA) (2006) Report to Congress on the Large Truck Crash Causation Study. MC-R/MC-RRA.

22. National Highway Traffic Safety Administration, *Pedestrian Safety*. www.nhtsa.gov/book/. Accessed March 9, 2024.

23. New York City Department of Health and Mental Hygiene (2017) Pedestrian Fatalities in New York City. March 2017, No. 86.

24. National Highway Traffic Safety Administration, *Motorcycle Safety* . www.nhtsa.gov/book/. Accessed March 9, 2024.

25. Johnson, Robert & Cureton, Adam, Kant's Moral Philosophy. *The Stanford Encyclopedia of Philosophy* (Fall 2022 Edition), Edward N. Zalta & Uri Nodelman (eds.). https://plato.stanford.edu/archives/fall2022/entries/kant-moral/

3 The Safety Generation

At the end of the 20th century, the U.S. Centers for Disease Control named the top ten great achievements of public health in the United States of the 20th century. Number three on the list was motor vehicle safety. Mid-century psychologists and physicians dealing with disease, injury and harm in the public sphere developed a conceptual approach to the epidemiology of accidents. This approach cast the road crash problem as an interaction between the human driver, the vehicle and the environment. Crash and injury statistics were collected and countermeasures that enjoyed the support of the federal government were developed. The success of this rigorous attack on road safety paved the way for motor vehicle safety regulation and led on to the new field of human factors research and eventually the new technology of Intelligent Transportation Systems (ITS) and the bold strategy of Vision Zero. Along the way, traffic engineers produced better designs for intersection safety, the design of the roadway infrastructure was oriented to safer highway operation and the insurance industry found ways to assess the crash risk of individuals. Means of encouraging safer driver behavior were developed, along with broad-scale enforcement of road rules and traffic law.

The road safety enterprise was able to grow rapidly aided by common cause made by medical professionals, engineers, economists, policy makers, law enforcement and many others. This era was highly productive from the 1960s through to the turn of the century – a simpler time before transportation's policy environment became more complex. It is so easy to get away from the raw material of safety – the injuries, deaths and destruction. Everyone wants to avert their eyes, even the safety analysts and experts. Some revert to abstractions and jargon, and a safety bureaucracy has grown up over the past 50–60 years. We have come a long way from the mid-1960s activism of William Haddon, Ralph Nader and others like them. Along the way safety has been homogenized and current generations may not know the unvarnished reality.

3.1 THE TALISMAN OF EPIDEMIOLOGY

In 1964, William Haddon, Jr. published the Haddon Matrix (1). While representing a relatively confined construction of the road safety problem, it was brilliantly clear

DOI: 10.1201/9781003483861-3

65

and inspiring to many. For the first time, the roles of vehicle driver or occupant (host), vehicle (agent) and environment (roadway) were called out as the columns of the matrix. The rows were the phases of pre-crash, crash and post-crash. The cells of the matrix contained pertinent factors that contributed to total harm. The epidemiologist knew that there were many possible factors, but tried to tie them to the part of the matrix where they had the greatest influence. Clear thinking about contributory factors could lead to a clear analysis of potential countermeasures. The epidemiological view of the vehicle as the agent, or the vector, of harm was surprisingly well accepted in the mid-century politics of public health, maybe because emergency room scenes provided graphic evidence of injuries caused directly by occupants crashing into vehicle interiors.

William Haddon's single-minded approach to motor vehicle crash injuries was by no means the obvious solution. But his matrix ended up being the talisman for continued safety improvements over many decades. The seeds were well-and-truly sown by the very nature of Haddon's stock in trade – epidemiology. Influential American political leaders of the 1950s and 1960s were alarmed when the true extent of road fatalities and injuries was revealed, even though it was not really an epidemic.

3.1.1 DRAINING THE SWAMP

In his devastating article entitled Epidemic on the Highways (2), published in The Reporter on April 30, 1959, Daniel P. Moynihan provided the following clarion call: "something more effective than simply urging people to stop killing each other must be done, probably through the intervention of the Federal government, to control this disastrous epidemic". William Haddon had convinced his former boss that only epidemiology – not engineering, nor medicine, nor economics, nor law enforcement – could control the rampant aspect of motor vehicle crash injury. The secret was to limit the exposure of the host – the driver – by changing his entire relationship with the agent – the vehicle – and the environment. This was tantamount to draining the swamp, rather than concentrating on the disease of malaria in itself. The Haddonites, and many others, were galvanized by the idea of making the vehicle a survivable space. They were not so enthusiastic about their prospects of improving driver behavior and avoiding crashes altogether – the equivalent of curing malaria.

However, as time passed, attention also needed to be directed to the host: the driver. Among the cognoscenti, this simple change of focus was accompanied by decades of in-fighting. Eventually, the great kitchen-sink economist Malcolm Gladwell weighed in on the side of the driver. In his 2005 *The New Yorker* article Wrong Turn (3), Gladwell castigated William Haddon and the Haddonites. Not so much for disregarding the driver, but for pushing airbags as primary occupant restraints, rather than seat belts. This subtle distinction was probably lost on many, but safety experts revealed themselves as true believers, one way or the other. Even though neither would avoid crashes, airbags would operate automatically and were therefore passive from the driver's perspective. Seat belts needed to be used, and used properly, so the driver had a crucial and active role to play – but would they? Haddon was all for the passive approach to safety.

Haddon didn't believe that a sufficient number of people would use seat belts on a sufficient number of occasions, while the airbag would be built into the vehicle and would deploy automatically. In Gladwell's conversations with Ralph Nader, he got the impression that Nader had never got beyond the civil liberties issues with the need for Americans to actually wear the seat belt. This presumption on the part of all of the Haddonites could be traced back to their association with New York Governor Daniel Patrick Moynihan in the 1950s. Haddon, Nader, Moynihan and Joan Claybrook, who succeeded Haddon as head of NHTSA, were all for airbags as the primary occupant restraint. This was part of their "drain the swamp" approach because they didn't think drivers could be trusted to always wear their seat belts.

3.1.2 Controversy over Haddon's Analogies

Gladwell's article also turns on the contention that, while the NHTSA era of road safety saved many lives and injuries, America did not become the world leader in road safety statistics. This distracting notion takes us even further away from the realities of crashes, deaths and injuries. Gladwell, and others like the safety professional Leonard Evans of General Motors, pointed to various formulations of crash data that showed better numbers for countries like Australia, Canada and the United Kingdom. However, to defend our man William Haddon we need look no further than the following excerpt from Gladwell's attempted take-down in *The New Yorker*:

> His goal was to reduce the injuries that accidents caused. In particular, he did not believe in safety measures that depended on changing the behavior of the driver, since he considered the driver unreliable, hard to educate, and prone to error. Haddon believed the best safety measures were passive. "He was a gentle man", Moynihan recalls. "Quiet, without being mum. He never forgot that what we were talking about were children with their heads smashed and broken bodies and dead people".

And the Haddonites got things done. By 1967 Haddon was installed as the first administrator of the National Traffic Highway Safety Administration. It is remarkable that Haddon's analogy between disease and motor vehicle injury had such an impact, that the mosquito that acted as the agent in the case of malaria could be credibly replaced by a motor vehicle. Despite a number of concerted attacks on Haddon's beliefs and methods, he and his organization stuck to "Plan A". They carried out methodical data collection and factorial analytics for all parts of the Haddon Matrix. They then turned this knowledge base into countermeasures, drove the implementation of their countermeasures and assessed their benefits. Again, Gladwell provides an eloquent testimony to the effectiveness of the Haddonites acting through the lens of NHTSA:

> There is no question that the improvements in auto design which Haddon and his disciples pushed for saved countless lives. They changed the way cars were built, and put safety on the national agenda.

If this was damning with faint praise, so be it. Haddon was not given to philosophical flourishes, but it is hard to imagine a more fitting encapsulation of beneficent public safety, utilitarian and fair application, egalitarian scope and legitimacy of authority. The rigor of epidemiology became the life force of road safety.

3.1.3 HADDON AND NHTSA

In 1966, William Haddon was appointed by President Lyndon B. Johnson to head up the National Highway Traffic Safety Administration (NHTSA), and he started it from scratch. Today NHTSA describes its mission as "to save lives, prevent injuries and reduce vehicle-related crashes". It develops and enforces Federal Motor Vehicle Safety Standards (FMVSS); it also has wide powers such as running the Fatality Analysis Reporting Scheme (FARS), administering the national Vehicle Identification Number (VIN) scheme, overseeing vehicle safety recalls and licensing vehicle importers. It also administers vehicle emissions and fuel economy regulations on behalf of the Environmental Protection Agency (EPA) (which lacks the regulatory muscle of NHTSA). Today, national vehicle safety regulators operate in many countries, and these organizations have similar scope and gravitas.

By 2020, NHTSA had more than 600 staff and an annual budget of $1 billion. And priorities had changed dramatically over the years, with driver safety now absorbing a much larger piece of the budget than vehicle safety. But NHTSA's original assault on the many physical dangers present in vehicle interiors had paid off and continued to pay off. Automakers were forced to provide a less aggressive, and more reliable, survival space within the vehicle, and to fit restraints in the form of seat belts, and eventually airbags. In fact, NHTSA's very first salvo of vehicle safety standards, enacted in 1968, included mandatory fitment of seat belts. But it was left to the states to require them to be worn, or not.

Most expenditure carried out by the U.S. Department of Transportation (DOT) is contained in multi-year "highway bills" that try to target the large sums to be found in the bill to the transportation priorities of the day. These periodic highway bills only devote a small proportion of their dollar allocation to safety, but important steps continued to be taken. The Intermodal Surface Transportation Efficiency Act of 1991 required all new cars to be fitted with airbags for the driver and front passenger, as of the 1998 model year. It had turned out that seat belts were a more natural driver restraint than airbags, which had to be aimed in a certain direction. After many years of refinement, airbags came to be viewed as important supplementary restraints, with a range of locations and purposes within the vehicle.

The only highway bill to actually mention safety in its title is the 2005 Safe, Accountable, Flexible, Efficient Transportation Equity Act: A Legacy for Users, known colloquially as SAFETEA-LU. As expressed in the 2005 announcement by the Federal Highway Administration (FHWA) – by far the largest of the U.S. DOT agencies that sit alongside NHTSA – the bill had a typically wide scope:

> SAFETEA-LU addresses the many challenges facing our transportation system today – challenges such as improving safety, reducing traffic congestion, improving efficiency in freight movement, increasing intermodal connectivity, and protecting the environment – as well as laying the groundwork for addressing future challenges. SAFETEA-LU promotes more efficient and

effective Federal surface transportation programs by focusing on transportation issues of national significance, while giving State and local transportation decision makers more flexibility for solving transportation problems in their communities.

The FHWA went on to address its safety programs as follows:

SAFETEA-LU establishes a new core Highway Safety Improvement Program that is structured and funded to make significant progress in reducing highway fatalities. It creates a positive agenda for increased safety on our highways by almost doubling the funds for infrastructure safety and requiring strategic highway safety planning, focusing on results. Other programs target specific areas of concern, such as work zones, older drivers, and pedestrians, including children walking to school, further reflect SAFETEA-LU's focus on safety.

This institutional safety commitment by NHTSA's large sister agency shows how far we have come from NHTSA's initial focus on the specifics of injuries and fatalities caused by crashes. In William Haddon's world, this 2005 expenditure was devoted to the "environment" column of his matrix, as well as certain neglected classes of the host – but not the agent. For Haddon, highway improvements were probably part of the swamp that needed draining but were still quite remote from the actual violence of crash injuries. The language of the bill also shows that we were no longer in epidemic control mode. And the crash statistics bore this out. The crisis of the 1950s and 1960s had been dealt with very effectively.

3.1.4 SAFETY IN THE OTHER U.S. DOT MODAL AGENCIES

The FHWA was created on October 15, 1966. Right from the start, this was easily the biggest U.S. DOT agency. Its history goes back to the Bureau of Public Roads in the early 20th century. It was designed to administer the federal aid highway program and was an important part of the age of engineering professionalism, spanning from the latter decades of the 19th century through to the mid-20th century. FHWA is responsible for preserving and improving the National Highway System (NHS), including the Interstate System. Its work is predicated on highly refined engineering standards. Under its current mission, "the FHWA works to improve highway safety and minimize traffic congestion on these and other key facilities". It is responsible for distributing the substantial proceeds of the federal gasoline tax – in the vicinity of $50 billion in 2020 – to state DOTs. This is the main source of funding for U.S. highway construction, maintenance, and operations.

The FHWA publishes the *Manual on Uniform Traffic Control Devices* (MUTCD), which is used by most highway agencies in the United States. The MUTCD provides such standards as the size, color and height of traffic signs, traffic signals and road surface markings. It is also responsible for comprehensive research programs, such as the Strategic Highway Research Program (SHRP2). Conducted in the early 21st century, this large naturalistic study addressed crash risks related to infrastructure factors. This was partly inspired by the many naturalistic driving studies carried out on behalf of NHTSA, studies that addressed the safety features of the vehicle. SHRP2 applied similar naturalistic methods to the roadway and provided valuable insight into crash risk and causation from a highway perspective.

The collective expertise of FHWA assists state DOTs to identify network locations of high risk for crash occurrence and to address such risks with appropriate countermeasures. Given the sheer size of the national road network and the longevity of roads and bridges – especially relative to motor vehicles – it is not possible to make dramatic improvements in roadway infrastructure safety. It is a slow, steady process, but significant in the long run – it all helps in "draining the swamp".

The geometric standard of a roadway has an important effect on the frequency of crashes that occur. This technical term could refer to lane widths, sight distances, curvature, emergency lanes and whether the roadway is divided or not. The presence of barriers and separation from roadside furniture is also important. Traffic streams that oppose or cross each other are clearly high-risk, but it is often difficult to interfere with other important functions of the built environment. Suffice it to say, that FHWA is extremely safety-conscious but is not an agency on the cutting edge of safety science. In recent years, it has adopted the safe system approach – that no one should die simply because they make an error on the road. But this ideal is still a long way from being achieved.

Safer roads are an important part of the safe system, and FHWA aims to "design roadway environments to mitigate human mistakes and account for injury tolerances, to encourage safer behaviors, and to facilitate safe travel by the most vulnerable users". Again, the safe system approach would not be a decent response to an epidemic. But it would definitely be part of the "safety-industrial complex" – as the current-day author Matthew B. Crawford called it (4) – that started to grow once the initial startling success of NHTSA was widely appreciated. Crawford has reservations about a partly commercial safety enterprise that operates at arm's length from the actual harm – the collisions and injuries. Suffice it to say that FHWA has left breakthroughs in car safety to NHTSA, and rightly so.

By virtue of their sheer size and weight, freight trucks represent the class of motor vehicle of most specific interest to FHWA. In terms of their impact on traffic flow, FHWA assigns an equivalence factor to various classes of vehicles. In this scheme, a car is assigned a Passenger Car Unit (PCU) of one. Depending on the grade of the roadway, a bus, heavy truck or large truck–trailer combination could have a PCU of up to six. That is, a large, heavy or slow-moving vehicle is assigned a take-up of roadway traffic capacity up to six times greater than a car. In addition to providing space in traffic lanes, roads are also designed to carry the high axle loads of freight vehicles, and FHWA applies truck axle and gross vehicle weight limits so that the design strength of the road pavement, or bridge, is not exceeded. Clearly, there could be safety consequences if major infrastructure failures – holes, fractures, separations etc. – are caused by heavy trucks. This happens infrequently and the road damage caused by trucks is seen more in surface cracks and ruts in the "wear and tear" category. Rather than a safety issue, this becomes a highway maintenance and consequent road use charging issue.

The main safety impact of heavy and large vehicles is caused by the inequality of the truck and car when they collide. This is related to gross imbalances of weight, size, shape and stiffness. Once the truck weight reaches a certain level – even without an actual payload being carried on the vehicle – further increases in weight have a diminishing influence on crash severity. Even if the driver of the truck is not held to be responsible for the crash, the severity of injuries in the car far exceeds any that occur in the truck. Imagine a tractor–trailer traveling at 60 mph, loaded to the federal weight limit of 80,000 lb. If a passenger car finds itself on a symmetric collision course at the same speed, several nightmares of physics happen very quickly. In less than a second, the car ceases traveling forward and begins traveling backward at a similar, but slightly reduced speed. The huge rate of deceleration involved in that velocity change of 110 mph immediately destroys any hopes of occupant protection or survival space in the car. And something else really bad happens. Because the tractor has a high-mounted and rigid chassis, the car is inexorably compressed downward and forced under the front of the tractor. The car becomes an unrecognizable billet of steel, and so much heat is generated as it scrapes along the road that it bursts into flames before the whole assembly gradually grinds to a halt.

Car-into-truck collisions involving the sides or rears of trucks are also extremely violent in ways that cars will never be able to resist. While it is generally not true that heavy trucks have higher crash involvement rates than cars, it is certainly true that crashes between cars and heavy trucks are much more severe and are much more likely to result in fatal injuries. For example, a passenger car impacting the rear end of a semi-trailer that does not have under-ride protection is going to be the instrument of decapitation for its front seat occupants. That is not to say that truck drivers are sitting pretty in a safe cocoon. Over-the-road tractors, and straight trucks, do not provide exceptional occupant protection. On the contrary, the truck driver is extremely vulnerable in certain types of crashes – especially those involving rollover – because truck cabins have very weak roof structures. There were no Haddonites to pressure truck manufacturers to suddenly start designing and building safer trucks. Many truck drivers have been killed instantly when their tractor lands off the road, upside down, and the cabin is crushed right down to the base of the windshield.

Is this a problem that falls into the bailiwick of the roadway and traffic people? In earlier times, road engineers believed that vehicles, including heavy trucks, should stay on the paved roadway. It was up to them not to leave the roadway. Anyone who studied heavy truck crashes in the 1980s – as I did – quickly learned that the actual locations coincided with piles of gravel on the roadside. The state road authorities quietly removed any roadway defects in the aftermath of fatal crashes involving trucks. This appeared to be done as a matter of policy, regardless of whether the road defect may have been a primary factor in causing the crash. In those days, the road shoulder on a major interstate highway could be narrow, and culverts could be extremely close to the edge of the traffic lane. The driver of a tractor–trailer combination that strayed into the culvert was likely to be crushed when the tractor overturned. The road authorities gradually assumed more responsibility for the role of the roadway in crash causation and severity but were not leaders in road safety.

On January 1, 2000, the Federal Motor Carrier Safety Administration (FMCSA) came into being as a truck and bus safety agency within the U.S. DOT. It has a mandate to reduce crashes, injuries and fatalities involving large trucks and buses. This 1,000-strong organization has a budget of less than B $1.0 and is focused on the vast number of carriers, drivers and trucks needed to carry out a massive transportation task: 70% of U.S. freight movement by value and 60% by weight. Considering that the United States is comparatively well equipped with competitive freight rail infrastructure, the dominance of road freight is indeed remarkable. There are over 1 million U.S. registered for-hire carriers. There are almost 40 million trucks registered for business purposes, including about 4 million Class 8 trucks – basically 80,000-lb 18-wheelers. About 3.5 million truck drivers are employed in the industry.

There have been 5,000 to 6,000 annual heavy truck fatalities over a number of decades, with noticeable temporary reductions caused by economic recessions. While truck fatal crash rates per unit distance traveled were higher than the rates for cars in the 20th century, that difference has largely disappeared in the new century. Fatal crash rates for heavy trucks and cars have generally followed the same long-term trend of reduction. FMCSA continues to exercise a high degree of surveillance over commercial vehicle crashes and expresses a desire to maintain a balance between the harm thus caused and the economically important, and massive, task of freight movement.

Some of the many instruments used by FMCSA include: carrier safety ratings, commercial drivers' licenses, driver medicals and hours of service rules. FMCSA maintains a consistent overview of crash data, with its own methods appropriate to the commercial – rather than personal – nature of truck driving. That data starts with NHTSA's Fatal Accident Reporting System (FARS) which is not a sample, but a census of all motor vehicle fatalities. This is supplemented with the General Estimates System (GES), which is a broad sample in that it includes fatal crashes along with the more numerous crashes of lesser severity, causing injury or property damage. Then FMCSA maintains its own census of truck and bus crashes – of all severities – and known as the Motor Carrier Management Information System (MCMIS) Crash File. The FMCSA encourages shippers to use carriers with good safety ratings. The carrier safety ratings have always been controversial and cover Compliance, Safety, Accountability (CSA) and the Safety Measurement System (SMS). The main purpose of SMS is to prioritize less-safe carriers for enforcement and vehicle inspection.

Hours of service rules have been assessed and adjusted repeatedly over the tenure of FMCSA. Such rules are complex in formulation and enforcement. In 2015, Congress directed FMCSA to carry out a naturalistic driving hours study of several hundred truck drivers over a five-month period. The methodology of this study started with the naturalistic study concepts pioneered by NHTSA over several decades but included several proprietary means of tracking driving, sleepiness and sleeping. The sensitivity of the subject of driving hours is reflected in the fine-grain variations investigated. Nevertheless, it was found that truck drivers were averaging less – maybe one hour less – than the widely accepted minimum of 7 hours per 24-hour period.

Continuing controversy is inevitable. FMCSA carries on doing its very difficult job, with small resources relative to the heft of the U.S. trucking industry. FMCSA is the place where the conflict between safety – propensity for crashes – and driving intensity – in this case, the rate of working by truck drivers – is in starkest relief. When does that work rate – when work happens to entail driving on public roads – become a danger to the broader community of road users? And are those other road users doing enough to protect themselves? An uneasy stasis continues.

3.2 THE NEW FIELD OF HUMAN FACTORS RESEARCH

The physical and cognitive aspects of the work carried out by human beings began to receive professional attention in the latter part of the 19th century. There was no organized field of activity until the military context of humans working efficiently, and humans working with machines, led the British Admiralty to convene a post-World-War-II meeting of professionals engaged in personnel research. At that meeting, Professor Hugh Murrell proposed the use of the term "ergonomics" on July 12, 1949. The British Ergonomics Society was created in 1952 and subsequently the U.S. Human Factors Society. The terms ergonomics and human factors have been used somewhat interchangeably for many decades, but the word ergonomics is generally used with reference to work situations, while human factors may refer to any interaction between man and machine. In the case of humans interacting with electronic systems, like computers, the term human factors would usually apply.

Today, both terms refer to the understanding of interactions between humans and other elements of systems, for the benefit of both the human element and also system performance. Human factors principles are essential to the design of complex mechanical and electronic systems, and the discipline has often concentrated on the man–machine interface. The field is notable for being highly interdisciplinary, bringing together engineering and applied psychology in particular. In retrospect, professional human factors, their knowledge and methods, seem to be invented for the automotive industry and road safety in particular.

Early in the 20th century, national associations of automotive engineers, like SAE Inc., began consolidating and standardizing knowledge of many aspects of the primary vehicle control task. The physical basics all needed to be created, including human anthropometry, reach and strength. The primary controls for acceleration, steering and braking needed to be invented and refined. Drivers needed to see where they were going, signal their intentions, learn to control the vehicle at night and remain informed about their motive power and speed. Consumers wanted vehicles that were faster but that would still be controllable by ordinary drivers, would not confuse and would not intimidate.

By the time human factors researchers directly turned their hands to safety – in the mid-1960s – the profession had already made huge contributions to automobiles' fitness for purpose and driver-friendliness. For the new era, there was a daunting range of questions being asked: How fast can drivers react? How far ahead of the vehicle do they need to look? How long can they look away from the road ahead? How do they track the roadway and judge their position in the lane? How should curves be designed and transitions from straight to curved? How much do they know about the behavior of other vehicles in their vicinity? How is the comfortable speed affected by roadway geometry? How should edicts and advisories be conveyed to the driver? How much information can they absorb, and how quickly? What is the right balance between responsiveness to driver control actions and stability? This is the sort of information needed to design a respectable vehicle: not only where the system performance avoid major pitfalls but also the driver would be able to cope and feel empowered. Respect for the human driver as part of the design was paramount.

3.2.1 NEW TECHNOLOGY ASSISTING HUMAN FACTORS RESEARCH

As this professional field advanced, and more data came to hand, researchers began to delve deeply into human behavior and to develop newer, grander research tools. Faster computers opened the door to driving simulators that could begin to address the many questions concerning the increasing number of controls in the vehicle, their appropriate sensitivity and their hierarchy from the driver's perspective. In addition to these many smaller, but important questions, researchers began to investigate the science of driving itself. Is it possible to model the driving control task with the human in the loop? Which motion cues are used to underpin the large number of motion scenarios that confront drivers? In addition to visual and vestibular cues, humans use the modality of proprioception – our ability to sense action – to achieve locomotion when the more obvious cues are not available.

Driving simulators were created at varying levels of verisimilitude. Unlike flight simulators, where the control task is less dynamic and less complex, driving simulators are not able to recreate automobile driving in all of its sensory, cognitive, control and response attributes. Visual aspects have become more realistic, with an enfolding field of view, and the dynamics of starting, stopping and turning may be created with a "moving base". But these two sets of cues are difficult to combine in a convincing way, and nausea is the result for many driving subjects inducted into driving experiments.

Human factors researchers have learned to live with such limitations, which blunt simulators use for studying the primary driving task. Rather, simulators have been used successfully for studying the effects of secondary tasks, such as cell phone use while driving. Driving simulator experiments conducted with human subjects have contributed greatly to our knowledge of human interactions with cars, especially the human–machine interface, field of view, information handling, reaction time, traffic controls and roadway innovations. Simulators have provided important indications of driver reactions in emergency situations. They have also been used extensively for driver training purposes.

Given the limitations of simulators in replicating the motions of a moving vehicle, researchers sought to transfer experiments inside modified motor cars. Such experiments would usually be carried out on closed test tracks. In the 1970s, the author carried out vehicle handling experiments with a "variable characteristic car" in which the driver's steering wheel motions were interrupted and contrived to reduce the stability margin of the vehicle. Early limitations of computers that could be readily installed in vehicles were quickly overcome, and it became possible to insert real-time simulation "in the loop" and to capture time history data from multiple sensors, including displacement, rotation and acceleration. Whether carried out in simulators or in special vehicles, experiments were designed to vary vehicle factors in a controlled way and to measure resulting changes in driver behavior. Because drivers' propensity for response varies greatly between individuals, the results would be averaged over a number of test subjects. The intent of such experiments was to arrive at vehicle control settings that led to optimum performance of the driver–vehicle system and to determine the settings that most drivers preferred.

At the same time, human factors researchers set out to determine roadway design parameters that were in accord with drivers' capabilities and desires. Roadway geometry, sight distance, lane markings, signage and signal locations were among the elements that were investigated. The research intended to maximize the transfer of essential information to drivers and to minimize the effort and attention required on the part of drivers. Much of this work was done in the controlled environments of dark tunnels, using projection and the like.

All of this research required means of measuring the "good" and the "bad" from the driver's perspective. Sometimes it was possible to directly measure the performance of the primary task, and sometimes it was necessary to run a secondary task. The transfer of attention from the secondary to the primary task would then show how demanding the primary task was proving to be. Physiological measures of task demand could be applied directly to the driver, to show how stressed they were. These included heart rate, galvanic skin response and blood pressure. There were several ways of measuring spare mental capacity. Other techniques applicable to the driver's visual performance used occlusion of the roadway scene, created mechanically. The frequency or duration of occlusion could impair driving performance and therefore point to the critical visual elements required by the driver. Various subjective means of measuring spare mental capacity were also developed and presented as specific scales that researchers could adopt; for example, the Subjective Workload Assessment Technique (SWAT).

3.2.2 THREE ERAS OF AUTOMOTIVE HUMAN FACTORS

In the early 20th century, the first era of automotive human factors – well before it was called human factors – was concerned with figuring out the norms of driver control mechanics, gauges, signals, adjustable seats, headlights and wipers. Early seat belts were developed, for reasons of comfort rather than safety, and the protective passenger cell was recognized. This was all done heuristically, almost by trial and error.

In its second era, from the 1960s, human factors research became more and more synonymous with the design of a safe vehicle. While early work on occupant protection in collisions was very dependent on its own scientific discipline, called biomechanics, the growing field of "active safety" was closely allied with human factors. There was a large appetite for research findings that would permit prioritization of the many factors that *could* be affecting safety. For example, the handling of vehicles could be too close to the edge of sudden control loss – as was suspected with the Corvair in the 1960s. Or, great care was needed to ensure a stable response to steering with high-center-of-gravity (COG) vehicles, like the 1980 Jeep CJ-5, which had a propensity to roll over. Glare from headlights could lead to control problems for the drivers of on-coming vehicles. The increasing volume of information destined for the driver could lead to mental overload or distraction from the primary task of driving.

All of this work continued to focus on making a vehicle as well-suited to drivers' capabilities as possible and without obvious vices. Along the way, the field of human factors became less concerned with physical systems and more concerned with cognitive systems.

A third era began in the 1990s with research into sensing driving risks, to warn drivers of adverse dynamics in their own vehicle, or of adverse motions of other vehicles. These technologies were fast and reliable enough to rival some of the dynamic processes carried out by human drivers. Rapid and accurate sensing, computation and control actuation opened the door to decisive control actions initiated by the machine, not the man. For the first time, such actions jumped in before the driver and did not make things worse – always better. Automakers introduced Electronic Stability Control (ESC) in 1990, and today virtually every vehicle on our roads is equipped with ESC. When ESC detects an excessive vehicle rate of rotation, it modulates the steering and brakes, and keeps paying attention, to counteract the loss of control. This could include braking one side of the vehicle more than the other side, if necessary. Drivers simply can't do this. In many cases, the driver is not even aware of the potentially dangerous event. However, much of the human factors work in the third era didn't try to bypass the driver, but warned the driver, or pushed the driver toward the needed control actions. Such "applications" – we were now getting totally into the digital age – rely on emulation of aspects of human driving.

3.2.3   A BETTER MAN–MACHINE SYSTEM

The field of human factors has accumulated an impressive basket of knowledge about the act of driving. However, this could not be considered to be a full science of driving and would not cover all of the aspects of sensing, planning and executing that human drivers routinely carry out. The wholistic modeling of driving did not come into focus until the 21st century when large-scale industrial R&D was applied to the development of Automated Driving Machines Systems (ADSs). Although the designers of ADSs use human driving as the touchstone of their work, they are not able to recreate driving in its entirety. Human factors research seeks to complement the person behind the wheel, not replace them. So all of the empirical and theoretical findings still leave many gaps in understanding the automotive version of a citizen or consumer. The many standards that apply to the physical and cognitive aspects of driving, and seek to harmonize machine and man, aim for a better system. This approach has long been predicated on the avoidance of mismatch that could prove difficult or harmful for the human in the system.

This benevolent approach, so characteristic of human factors research, includes awareness of its limitations. The machine part of the man–machine system is usually designed with a margin of safety. This is intended to allow for hidden deficiencies in our understanding of driving phenomena, as well as the immense variability between drivers. And, recognizing that people are definitely not machines, human factors experts have also borrowed extensively from the field of applied psychology.

3.3   MORE AND BIGGER ROADS – AND SAFER ROADS

The aeronautical age of the early 20th century captured the imagination not only of the public but also of engineering designers everywhere. The obvious connection between form and function in reducing drag – the enemy of all machines that seek to travel fast, even in air – created a new fashion of aerodynamic design. Streamlined vehicles were immediately recognizable and spoke for themselves in terms of advanced design and performance. They became the very epitome of modernity. In the case of cars traveling on roadways, it wasn't too much of a stretch to then see a stream of traffic akin to a fluid moving through a pipe. This extremely modern idea made travel seem as simple as turning on a tap and as benignly expansive as a telephone network.

These concepts were magnificently packaged by Norman Bel Geddes, an extraordinary combination of theater designer and industrial designer. Having designed sets for the Metropolitan Opera and Cecil B De Mille, he announced himself as an industrial designer with his 1932 book Horizons. This book popularized streamlining as a far-reaching style of design for many types of products. Born in Adrian, Michigan in 1893, Geddes was no stranger to the growing Detroit auto industry. He was hired by General Motors to design their famous pavilion at the 1939/40 New York World's Fair. His Futurama ride was a runaway hit.

According to Richard Wurts' well-illustrated records (5), a 36,000-square-foot model of the "highway world of 1960" was laid out. This included seven-lane roads with design speeds of 100 mph, experimental homes, farms, urban developments, industrial plants, dams and bridges. It included traffic streams flowing along separated carriageways and new types of intersections between intersecting roadways. The viewer levitated across this vista in one of 600 chairs, each fitted with its own loudspeaker. The entire production was at once outrageous and strangely normal.

Futurama was by no means an incongruous event. The 1939 fair was a testament to man's ability to command the physical world, produce highly desirable goods as never before, be entertained and inspired on a large scale and especially to create customized futures. The ubiquitous use of electricity was both a metaphor for an unlimited future, as well as a promotion of commercially available electric appliances that were still under-appreciated by consumers. The Transportation Zone included the Aviation Zone, Railroad Building, Chrysler Motors Building, Firestone Building, Ford Building, Goodrich Building and of course the General Motors Complex. In contrast to General Motors, the railroad industry featured its glorious past and the tire companies emphasized the durability and safety of their products; of course, it is not so long since many a car crash was blamed on tire failure, rightly or wrongly. It is remarkable that General Motors captured the world's imagination with their depiction of futuristic roadways and endless traffic streams, rather than the automobiles themselves.

As Wurts noted, streams of cars were still considered to be a good thing – even something to aspire to – back in 1939. Above all Futurama simulated a flight across the America of 1960 – looking 21 years into the future. It showed expressways – curving, bifurcating and full of cars and trucks – traversing bucolic rural areas and infiltrating skyscraper-dominated downtowns. Not everyone bought into this vision. The national columnist and architecture critic Lewis Mumford offered that "Mr. Geddes is a great magician, and he makes the carrot in the goldfish bowl look like a real goldfish... ..the future as presented here is old enough to be somebody's grandfather". However, life did imitate art and at least some of us still find the endless magic carpet of Los Angeles at night, with its illuminated arteries, its meta-traffic, a curiously satisfying representation of life itself. Before Futurama participants vacated their floating chairs, the last thing they heard from the popular newsreader who had served as their narrator was:

> With the fast, safely designed highways of 1960, the slogan "see America first" has taken on a new meaning and importance. The thrilling scenic feasts of a great and beautiful country may now be explored, even on limited vacation schedules.

Note that the important caveat of safety was included. You wouldn't just travel fast and far, but you would do it safely. It was also implied that you would do it effortlessly and enjoy the ride. You would be flying – at ground level.

3.3.1 VISION OF MODERNITY

But General Motors was alone in attempting to show its products as part of a seamless vision of modernity. Others were still content to drive home the much less ambitious message of overwhelming professionalism and command of their particular craft. The Ford exhibit deconstructed the manufacturing of a car, from the extraction of the required minerals through to the completed vehicle. The Ford Cycle of Production stopped short of envisaging what to do with a whole lot of cars – how they would move and interact with one another. The highlight of the rail exhibit was a streamlined steam locomotive weighing 526 tons and operating at 60 mph on a bed of rollers. The B.F. Goodrich pavilion included a 90-ft tower housing a massive tire guillotine – demonstrating the way tires were tested for strength and durability. The massive locomotive embodied an industrial world on the cusp of a thrilling transformation – futuristically aerodynamic in appearance, but still powered by steam.

One could be forgiven for thinking that General Motors would go back to Detroit and start designing and building roadways. This definitely did not happen. Many years earlier, it was Henry Ford who had rejected the idea of the private auto industry funding road construction, arguing that states and counties should have that responsibility. However, roadways as ambitious as those depicted by Geddes in 1939 would require a higher calling – the federal government. Visionary ideas for constructing nation-connecting roads had been circulating for decades. Leaders of the Bureau of Public Roads, and associations promoting regional routes like the Lincoln Highway, had produced a number of plans, including the Pershing Map, and culminating in a 1939 report (6) by Herbert S. Fairbank entitled Toll Roads and Free Roads. This was the first outline of the U.S. interstate highway system. Congress finally acted in 1944 with the Federal-Aid Highway Act:

There shall be designated within the continental United States a National System of Interstate Highways ... so located as to connect by routes, as direct as practicable, the principal metropolitan areas, cities, and industrial centers, to serve the national defense, and to connect at suitable border points... .

While the long-term federal roads supremo Thomas MacDonald – probably turned off by Norman Bel Geddes' showmanship – had been lukewarm about Futurama, he recognized that the public needed to be sold on big roads. The long lines of Americans waiting patiently, day after day, to get into Futurama showed that the time might be right. But where was the substance needed to go beyond the inspiring concept and design direct connections in the form of flowing roadways without distractions such as local commerce? At least some ideas had previously surfaced in the form of Fritz Malcher's 1935 Steadyflow System (7). Even earlier a Harvard researcher named Miller McClintock (8) had developed certain theories based on the fluid analogy of traffic movement. He believed that traffic accidents were caused by frictions in the flow – at intersections, in the median, within the flow and at the margins of the roadway.

3.3.2 TOWNLESS HIGHWAY

Some early traffic experts referred to the "townless highway", giving birth to the enduring idea of the hopefully frictionless limited-access expressway. In addition to the elimination of commercial distractions along the edge of the roadway, the key safety innovation was the divided highway, avoiding the risks of opposing traffic streams on the same carriageway. The idea of the townless highway was based on the assumption that big roads would be used for long-distance travel rather than local commuting and the like.

Prior to 1939, superhighways built for longer distances and higher speeds were being promoted, but there were nagging concerns about safety. Research on the effects of rapid increases in vehicle power and speed, and the continued adequacy of roadway design, had been initiated a few years earlier. The 1934 highway bill allocated 1.5% of each state's federal highway aid to highway planning and research. Thomas MacDonald oversaw a series of surveys that provided an enduring basis for highway scholarship. Forty-six states participated in mapping roadway hazards like tight curves, ditches, exposed culverts and narrow bridges. Enquiries also extended to the effects of variables such as curves, grades, lane widths and sight distances on behavioral impacts such as speed adjustments and positioning in the lane. The idea was to prioritize road upgrades from a new perspective of safety. As these studies progressed, the work became more intensive in probing unknown driver skills, including overtaking, passing and grade negotiation by larger and heavier vehicles.

This was the real world that Norman Bel Geddes' diorama soared above. But the frictionless, streamlined traffic flows postulated for 1960 were not entirely a fantasy. The sweeping, multi-lane high-speed expressway would overcome many of the safety problems that MacDonald's researchers had unearthed during the 1930s. This was a prerequisite of new roads built for effortless distance and speed. And

sure enough, long experience has shown that – on the basis of crashes per mile traveled – these higher standard highway facilities are at least twice as safe as the older
highways. However, none of this occurred with anything resembling a blueprint for
safety. While the benefits of reducing "friction" in the traffic flow – especially from
opposing traffic – were fairly obvious, many unforeseen ripples were created in the
small, placid pond of traffic engineering. For example, MacDonald's collaborator
Fairbank found that the townless highway was a very thin reality because 88% of trips
covered less than 30 miles. Rational engineers to a fault, MacDonald and Fairbank
then took to advocating free-flowing highways in the cities. Even though these were
safer in themselves, their superior performance on many other fronts became obvious,
and they encouraged more driving that then eroded overall safety benefits. Another
example of greater "driving intensity" being encouraged by the creation of a safer
driving environment.

3.3.3 EISENHOWER'S NETWORK OF EXPRESSWAYS

As engineers and entrepreneurs did the hard, detailed work to move beyond baby
steps with the design and operation of such facilities, others were more worried
about the politics. Dwight D. Eisenhower had long been attracted to the military uses
of an extensive highway system for large-scale movements within the country and
out to multiple points on its borders. Others saw a new avenue for mobility within
cities, using expressways rather than conventional city arterial roads. Could federal money be used for such expressways? And would city expressways affect the
quality of life in inner urban areas? The resounding answer was yes in both cases.
Such huge and socially alienating constructions in virtually all of America's big
cities caused definitive urban blight and social unrest. The long-term social impacts,
including health, employment and security, affected entire neighborhoods and lasted
for generations.

Big roads depend on big ideas, and big ideas can change remarkably quickly,
and can prove to be totally wrong. Effortless nationwide sightseeing was a big
idea in 1939 and maybe in 1960. But how would we cope today without the economic magic carpet of the interstate system and the urban conveyor represented by
the city expressway? And a lot of progress was made safety-wise. We are indeed
able to travel far – very far – and fast, comfortably and safely. Nobody could have
imagined just how large a population would drive cars freely, how many trips they
would make and how often. While this does equate to incredibly long aggregate
distances driven, these trips proved not to be "far": the preponderance of driving
remained local.

3.4 VEHICLE SAFETY REGULATION

The creation of NHTSA and the U.S. DOT came on the back of decades of engineering professionalism, standardization and the application of economics in professional public service. In the first half of the 20th century, the ideals of the engineer
rose to prominence. The standards movement had been underway since the 19th
century. The Federal Roads Act (1916–1940) and the creation of the Interstate
Highway System were based on faith in professionals and their ability to bring about

improvement. This was John Stuart Mill's progressive principle in action. Mill said that this principle could be expressed either as the love of liberty or improvement (9). The rise of democratic forms of government all over the world meant that the emphasis could be placed more on improvement. The 19th-century philosophers were very conscious that improvement required a desire and ability to rise above mediocrity and the "magical influence of custom". How could one indulge in standardization, somewhat bleak economic analysis, and even regulation, and still be inspired to improve the commons and all of its constructions and operations?

By the mid-20th century, there was an accepted need for regulating motor vehicles so that occupants were not injured gratuitously in the crashes that were occurring all over the country at an increasing rate. Nobody needed convincing that continuing mischief was certain unless new and definitive steps were taken. And the target of the new vehicle safety regulations – the automotive industry – had already been painted as passive and even deceptive when it came to crash protection. The creation of NHTSA in the 1960s is considered to be one of several landmarks in the pantheon of federal regulatory history, which includes giant steps such as: the Federal Trade Commission (1914), the Securities and Exchange Commission (1930s), the EPA (1970) and the Department of Homeland Security (2000s).

3.4.1 Federal Rulemaking

Federal regulations are sets of requirements issued by a federal agency in order to implement laws passed by Congress. Rulemaking is the process used by executive branch agencies to issue new federal, state or local regulations. The Code of Federal Regulations increased from 23,000 pages in 1960 to 180,000 pages in 2019. According to the Policy Circle (10), the cost of regulations increased 20-fold from 1960 to 2020. Virtually every industry is regulated, and the most regulated are those, like energy and consumer manufacturing, that deal most directly with consumers. The regulatory state is definitely alive and well and reaches into every corner of our lives. It is estimated that the total cost of all regulations is roughly equivalent to aggregate income tax receipts, or total federal government expenditure.

Motor vehicle safety was an early adopter in the age of regulation, and many other forms of regulation have emerged over the intervening decades. Nevertheless, motor vehicles are still the third-most regulated sector at the national level, after fossil fuels and electricity. By comparison, drivers and roadways are much more lightly regulated. While the Manual of Uniform Traffic Control Devices (MUTCD) was launched in the same mid-century time frame as NHTSA, it applies to a large government sector of much smaller economic significance. Model traffic laws were developed in the 20th century, and each state then came up with its own traffic law. These state laws have core similarities but are considered to be the responsibility of states in the 21st century.

3.4.2 Federal Motor Vehicle Safety Standards

The cornerstone of vehicle safety regulation by NHTSA is the set of mandatory standards known as the FMVSS. Manufacturers are permitted to self-certify the compliance of each vehicle model through tightly specified tests. But there are forbidding monetary penalties for non-compliance. In tune with Haddon's three-level scheme for injury prevention, FMVSS contains three series of rules, described as: crash avoidance, crashworthiness and post-crash survivability. In order to become part of FMVSS, these rules need to be well-defined, testable, verifiable in their ability to reduce deaths and injuries, and sufficiently long-lived. Vehicle safety rules were energetic in addressing early targets of occupant protection in interior impact, braking and lighting. Perhaps the most famous rules are Seat Belts and ESC, each emblematic of the major classes of vehicle safety rules: occupant protection and crash avoidance respectively. In recent years, fewer and fewer new rules have been produced. This is partly because software now plays a larger role in vehicle safety systems and cannot be regulated in the same way as the earlier waves of FMVSS.

As awareness of vehicle safety has expanded dramatically, and a worldwide ecosystem of expert private, public and academic entities has proliferated, it has no longer been necessary for NHTSA to always start the ball rolling with new rules or to remain eternally vigilant in enforcement. For example, in October 1997, a Swedish journalist carried out a "moose test" on a Mercedes Benz A Class passenger car. This test was intended to gauge the ability of the vehicle to avoid colliding with a suddenly appearing obstacle, such as a moose in the roadway. The test involved moving a certain distance sideways while traveling at a speed of 78 km/h, or 48 mph. The vehicle grossly over-reacted to the sudden steering required in the test and rolled over, sounding alarm bells in vehicle safety circles around the world (11). Mercedes Benz promptly upgraded the suspension and tires fitted to new A-Class vehicles. They also added technology known as ESC to the entire A Class. ESC uses a combination of vehicle motion sensors, actuators and software to sense when the vehicle is on the edge of losing control and to make virtually instant, perfectly calibrated corrections. These corrections mainly involve strong braking of a particular wheel, in order to prevent rapid yawing and rollover. The driver plays no role in this process, of which they are barely aware.

This singular sequence of events brought wide attention to the effectiveness of ESC in avoiding crashes and to the fact that it was applied to entry-level models in the Mercedes range, rather than the more expensive models. By the end of 2009, Ford and Toyota announced that all of their North American vehicles would be fitted with ESC. Soon after, NHTSA used the FMVSS system to require all passenger vehicles to have ESC. NHTSA's ongoing research showed that many thousands of lives were saved with ESC, and it is regarded as one of the single most effective vehicle safety technologies. But the ESC horse had well and truly left the barn before NHTSA started enforcing its ESC rule. The tried-and-true NHTSA process sequence of research, consultation, evaluation, public comment, promulgation and certification was effective and fair for hardware such as braking systems, but is far too linear and ponderous for software. Fortunately, the excellence of the innovative ESC process has created ongoing momentum and appetite for vehicle safety improvement via FMVSS.

3.4.3 Beyond Minimum Standards

On May 21, 1979 the first frontal crash test conducted at a speed above 30 mph was carried out. This was the first of many actions that expanded safety science beyond the strictly defined range of testing required in FMVSS. It was always understood that FMVSS was a minimum standard from the safety perspective because foreign trade could not be compromised. That same year, 1979, NHTSA established the New Car Assessment Program (NCAP). While the FMVSS was almost invisible to the consumer, NCAP was designed to give car buyers solid but understandable safety information on the new car they were about to buy. Well known for providing crash ratings for all makes and models of car, NCAP has been expanded over the years to include crash avoidance technologies like lane-keeping and blind-spot monitoring.

There is no doubt that nationally uniform vehicle safety regulation, and consumer information, stands out as one of the major achievements of the current regulatory age. As part of its regulatory process, the federal government has established automotive companies as regulated entities. But there is much more to road safety than vehicles. And roadways are not regulated in the same way. It is widely accepted that roads are the responsibility of governments, not the private sector. According to the Bureau of Economic Analysis, highways and streets represent the largest category of fixed assets, and 98% are owned by state and local governments. To many, topdown plans for highway infrastructure amount to federal intervention in investment decisions taken more appropriately by state, local and private entities. While federal rules are applied to limit the environmental impact of road projects, federal impact on highway safety – applied via the civil engineering of the infrastructure itself – has been minimal.

Even though road safety has long been recognized as a complicated dance involving drivers, vehicles and the environment in which they operate, regulatory power has been directed mainly to the private sector and a single product – the vehicle – and those who design and manufacture it. In addition to the stunning success of midcentury activists in targeting auto manufacturers for the long haul on safety, fuel economy and emissions were added to the vehicle regulators' remit. This compelling package of societal concerns works well for federal regulators because it is applied to the original equipment. Once a vehicle has been so built and certified, its intrinsic merits continue to deliver throughout its useful life, within reason. Federal regulation of motor vehicles for safety and emissions therefore makes perfect sense, and that regulatory power has been maintained unchallenged for many decades. This form of motor vehicle regulation represents the most extensive case of government influence over a consumer product.

3.4.4 CONTINUED FOCUS ON THE VEHICLE

Of the three elements in the Haddon Matrix, only the vehicle resides in the private sector and is of course extensively regulated by the federal government. The environment element, closely identified with the physical road system and its appurtenances, is owned by governments – state governments – and is subject to a wide range of standards. Standards on Interstate Highways are regulated collectively by the state highway departments under the American Association of State Highway and Transportation Officials (AASHTO). However, there is no concerted, scientifically informed effort to transform the road network from a safety point of view. Such a legacy system does not lend itself to systematic reduction of traffic conflicts, for example. The third Haddon element – the driver – is subject to state-level traffic laws and monetary penalties for breaches. Again, there are few signs of transformational changes in the way drivers drive, although increasing surveillance in some jurisdictions will bring breaches to light at a greater rate. The long-standing influence of the federal government over automotive companies in vehicle safety stands in contrast to their reluctance to interfere with the private behavior of individuals.

Another important form of federal regulation – economic regulation – also broadly impacts the automotive industry. It is aimed at consumer protections and stable markets. But there is no direct relationship between economic regulation and road safety. Economic factors definitely impact safety – for example, if younger, riskier drivers are forced to own cheaper vehicles with lesser safety features. But regulators have not used economic regulation as a lever to improve vehicle safety directly. If we look hard, we can find certain very limited exemptions from sales tax for virtuous products, such as exhaust emissions test equipment, but there is little evidence of economic incentives to specify or add safety features in vehicles. However, money does affect vehicle safety in practice. In recent decades, the more advanced driver assistance systems (ADAS) have been associated with luxury and high-end vehicle models. Such features could either be included as original equipment or available as options. Usually these options are offered in certain packages by automotive marketers, and it is up to the vehicle manufacturer to decide how to package such safety technologies and offer them to the market. In automotive insurance markets, vehicle insurers may provide some safety benefit by offering certain discounts for the presence of safety features.

3.4.5   REGULATION OF HEAVY VEHICLES

Economic regulation of large and heavy freight vehicles is a separate matter. Such regulation – for example, banning the haulage of certain commodities on certain roadways, in certain tunnels and on certain bridges, at certain times – is relevant to safety. The risk of fires in tunnels with flammable loads is an obvious case in point. The massive number of freight trucks operated by America's for-profit carriers and private fleets are subject to federal and state size-and-weight regulations. In a highly competitive and low margin industry, carriers attempt to maximize their payloads in terms of weight and/or volume on each vehicle. Devices designed to lessen the severity of collisions between these trucks and passenger cars are purchased and fitted by some fleets. However, many carriers may be turned off by the additional weight of devices such as under-run protection and the reduction of payload.

The use of driver assistance and crash avoidance technologies in carriers' fleets is limited. A number of FMVSS rules pioneered for passenger cars are – in principle – applicable to heavy trucks; but these requirements have been slow to trickle down to over-the-road freight vehicles. Professional drivers tend to have strong views about driver assistance devices and may resist their use. This creates great uncertainty about the wisdom of such investments in commercial fleets.

3.4.6 ETHICS OF VEHICLE SAFETY REGULATION

The first wave of safety regulation was applied to food and drugs and occurred in the early 20th century. A second wave occurred mid-century with occupational health and safety, environmental protection and consumer product safety. Health and safety regulation, which includes vehicle safety, raises important matters of ethics. While our prime objective is to protect the public from harm, this needs to be done in such a way that democratic virtues are preserved: these ethical dimensions include freedom of choice, consistency and participation in decisions. It may even be true that regulation is a substitute for morality in areas where our moral resources are lacking. And uncertainty and delays dilute the ethical imperative of safety regulation. The principle of "justice delayed is justice denied" applies. And it takes a lot longer to improve the physical infrastructure than to improve a cadre of vehicles. It is only with the advent of Vision Zero in the 21st century that road and traffic designers are receiving more scrutiny. And the rulers of the infrastructure should make new accommodations for ITS on the basis of the rapid deployment horizons of the technology. ITS is an accelerative safety avenue for infrastructure frozen in time.

Road safety regulation that focuses almost entirely on the vehicle has proven to be highly effective but is of course an incomplete solution. Vehicle safety rules fulfill their purpose of protecting consumers and the public. NHTSA's innovative and sustained efforts have borne great fruit, but the corresponding state of road and traffic standards and traffic rules has lagged. We are again reminded of the utilitarian nature of road safety. The consequences of collisions are frozen in physical evidence – mostly in crashed vehicles – and vehicle countermeasures almost announce themselves. By comparison, the rules by which drivers drive – and environments in which they drive – are general and are not specific to the avoidance of crashes.

3.5 TRAFFIC ENGINEERING

In the 2006 blockbuster movie Mission Impossible III, the actor Tom Cruise plays an undercover agent of international mystery. He doesn't want anyone to know how exciting his life really is. There is a scene in a bar where he is asked by several attractive people about his occupation. He says he's a traffic engineer with the state DOT and enthuses briefly about studying ripple effects in traffic. As he moves away, his interlocutors are seen yawning and smirking. The message is pretty clear: traffic engineering is incredibly boring and the perfect cover for someone of great daring, imagination and charisma.

The profession of traffic engineering developed largely in the public sector and these days every city and county needs a traffic engineering department, plus access to expertise in large traffic consulting firms. Particularly in the 20th century, the state DOTs developed traffic engineering solutions to deal with ever-increasing traffic levels and to accommodate greater vehicle performance capabilities. The state DOTs are supported by the U.S. Department of Transport's FHWA. FHWA is the largest of the U.S. DOT modal administrations and is responsible for national highway engineering standards and the allocation of substantial funding for the execution of those standards. Importantly, it is entirely responsible for the NHS, which includes the Interstate Highway System.

3.5.1 FOCUS ON HIGHWAY OPERATIONS

Obviously the design, construction and maintenance of highways have advanced steadily over the years. But in recent decades the biggest improvements have been in what the DOTs call operations. How do you get the greatest throughput of traffic, and the least congestion, on a given section of highway infrastructure of finite physical size and fixed disposition relative to the prevailing terrain? To address this question, standards and methods for traffic control have assumed great importance. Since the 1990s, a new branch of methods and technologies called ITS has arisen in the knowledge that improvements in the subtle dynamics of cars, roads and drivers may need new means of sensing and connecting all these pieces together. At last, the 1930s highway visionary's idea of frictionless, safe traffic flow may be coming into view. Though it was only included by implication, such traffic flow would also be crashless. Given the prevailing foibles of drivers, engineering means of bringing this about were unknown at that time.

If you had been able to corner Tom Cruise's character in that bar, he would have told you that traffic engineers are motivated to improve the safety and efficiency of highways, roads and streets. But unlike infrastructure or vehicle engineers – who actually design and build tangible things – traffic engineers are more virtual. And traffic engineers and ITS practitioners have it pretty tough. Unlike physical infrastructure, traffic and ITS standards evolve quickly and change because they interact directly with drivers. And driving changes in the same way that drivers' behavior is regularly changing. We should not assume that traffic engineers don't have new ideas but elected officials have a low tolerance for experimentation – trial and error – in public. Nothing says public mis-administration like long lines of cars, detours, the appearance of un-synchronized traffic signals, misguided priority at intersections and the like. Traffic jams are synonymous with world-weariness and pessimism.

Traffic engineers are risk-averse with very good reason. They take their responsibilities for safety seriously. At-grade intersections of traffic necessitated by roadways that cross over each other need rules and controls. We still rely on the century-old technology of traffic lights – red, green and amber. Many intersections throughout the country that should have traffic lights still do not have traffic lights. Those that do have traffic signals are not always maintained. Traffic signal manufacturers rely on a little-known device called the safety monitor. This device is standard in all traffic signal controllers and makes absolutely sure that a green cannot be displayed to two conflicting streams of traffic at the same time. Beyond this sacred safety commitment, many improvements have occurred over the years, making the phase changes smarter and adaptive to levels of traffic in the different directions. With multi-lane intersections, turning movements and pedestrians crossing, intersection design became an extremely specialized part of the industry.

A sobering number of fatalities and injuries continue to occur at signalized intersections and at intersections where stop and caution signs are relied upon. Even signalized intersections require compliance on the part of the driver. Given the relentless flow of high-density traffic in many places, lapses that are infrequent for any given driver add up to a considerable number of crashes in the weekly and monthly life of a busy intersection. You could say that these problems are place-based rather than people-based.

3.5.2 POOR COMPLIANCE WITH TRAFFIC CONTROLS

Stopping at red lights is obviously part of every state's traffic code and is enforceable by traffic police. But complete compliance on the part of most drivers is mixed. Some will continue through the intersection right on the cusp of the phase change to red and after the signals change. Some will speed up to reduce the amount of red time they claim for themselves. Given the limits of police enforcement out on the road system, many drivers are pretty sure they can get away with a little cheating at traffic signals. An automated antidote – installation of red light cameras – is simply too intrusive for many voters.

Traffic engineers have a certain amount to say about speed limits, but others including police and safety agencies also play a role in setting speed limits. This mix of responsibility varies from one jurisdiction to another. When roadways are initially constructed, they have a certain design speed and this determines permissible curvature, the width and shape of the cross-section, sight distance and grade. Because roadways have a very long life – certainly relative to the cars that drive on them – they tend to be re-purposed from time to time without being replaced. For example, lanes could be narrowed and more lanes created. Speed limits are influenced by the design speed, but other factors come into play. In many places, speed limits are based on the speeds that users tend to be comfortable with. But speed measurements will usually show a significant spread in the speeds that users choose. The speed limit is supposed to be set so that the vast majority (say 85%) of users choose an operating speed at or below the designated number to be placed on the sign.

In a few jurisdictions, speed limits are set by the state DOT and are based on safety criteria. In such jurisdictions, the criteria tend not to be absolute and posted speed limits may be lowered over time. After all, once a crash occurs, it is difficult to argue that a lower speed will not reduce the risk of injury. In cases where speed limits are set below the comfortable speed for large numbers of drivers, speed enforcement becomes an issue and speed cameras begin to be seen on roadsides. This type of automatic enforcement technology, like its cousin the red light camera, is often controversial. Although it is often viewed as an affront to personal liberty, some would argue that reductions in injuries and fatalities justify its use.

3.5.3 LIMITS OF TRAFFIC ENGINEERING

Throughout the latter part of the 20th century, road authorities became more and more aware of the violent injuries that occur when vehicles traveling at speed collide with fixed roadside objects like trees, utility poles, bridge pylons or with engineering structures like embankments and culverts. Just as safety epidemiologists saw the importance of vehicle occupants staying in the vehicle, highway engineers realized that vehicles need to remain on the roadway. Various types of barriers have been deployed widely – but still not widely enough – to redirect errant vehicles, absorb impact energy and reduce vehicle speeds at impact. This process occurred over decades as attitudes in the highway profession softened toward drivers who lose control of their vehicles. Not everyone had embraced the idea that drivers who make such errors and are demonstrably "in the wrong", deserve protective engineering solutions. These crash barriers and the like cost money; should that money be directed to help more blameless parties?

These days safety is in everyone's mission statement, but many involved in contributory activities like traffic engineering lack real first-hand understanding of the harm that goes with traffic injuries and deaths. We have seen major improvements in vehicle safety, brought about by the protection of occupants in the first instance and more recently through the avoidance of crashes altogether. Highway safety has tried to follow a similar pattern by placing a cordon between vehicles moving uncontrollably and passive objects that may then present themselves to vehicles in crashes. There is a certain analogy between the unrestrained vehicle occupant being injured by the vehicle interior and the out-of-control vehicle being injured by fixed roadside objects.

3.5.4 POTENTIAL OF ITS

However, the idea that traffic engineers could bring about a revolution in crash avoidance has not taken root. If an intersection has modern traffic controls, and perhaps is equipped with red light cameras, what more could the traffic engineer do to prevent the occurrence of crashes? In recent years, ITS technology has been trying to fill that gap. Continuous transfer of intelligence between such an intersection and vehicles in the vicinity could certainly provide drivers with much more accurate information about imminent crash risks. This type of wireless communication, combined with real time detection of precursors to known crash types, is a paradigm called "connected vehicles and infrastructure". Its well-recognized technologies are termed Vehicle to Vehicle (V2V) and Vehicle to Infrastructure (V2I) communication.

V2V and V2I (V2X) do provide new ways of avoiding crashes. In the case of a vehicle running a red light, "innocent" vehicles traveling with a green in the cross direction may be warned. In a different example, a sudden stoppage on a busy freeway, approaching vehicles moving at free speed can be warned and avoid the risks of panic stopping. Extensive accumulated knowledge of crash scenarios that occur over and over again can be put to work to design use cases, or applications, that lend themselves to cancellation with V2X. Unlike vehicle and infrastructure engineers, ITS engineers are all-seeing and all-knowing about the creation of crashes but lack empowerment and get lost among the big silos of large government organizations.

Vehicle manufacturers have a moral, as well as legal, responsibility to eliminate human injuries that occur inside their vehicles. In a similar way, infrastructure builders have a moral chain of responsibility to eliminate vehicle smashes that occur on their facilities. Thus far, infrastructure builders have failed to fully address these responsibilities – perhaps because they are part of the government and more heat has been applied elsewhere. But with advent of connected infrastructure the necessary knowledge for crash cancellation resides within their very organizations. However, this precious expertise in ITS and traffic engineering tends to be pushed to the fringes and so far struggles to enact meaningful change.

3.5.5 SAFETY IN STATE DOTs

There are 50 state DOTs, large and small. And there are significant differences between the structures of state governments and exactly where state DOTs fit in. In the case of the large State of California, there are many transportation-related agencies under an umbrella organization called the California State Transportation Agency (CalSTA). The DOT is known as Caltrans and sits alongside the Department of Motor Vehicle (DMV) and the California Highway Patrol (CHP). Also in this group is the Office of Traffic Safety (OTS). OTS oversees campaigns, programs and grants focused on well-recognized safety issues, including alcohol impairment, driver distraction and emergency medical services. In addition, the OTS provides increased attention to previously neglected modes of travel, including motorcycle riding, cycling and walking.

Beyond the shop-window of OTS, all the other agencies have safety activities behind the scenes. For example, the major functions of Caltrans are:

- Operations (including both the Office of Safety and the Office of Mobility);
- Maintenance;
- Planning (including freight);
- Design (including changes to design standards);
- Local Assistance (supporting cities and counties with guidance); and
- Programming (factoring all of Caltrans' projects into the funding process).

The safety imperatives of traffic engineers and ITS specialists are contained in Caltrans' stewardship of traffic control devices, including traffic signals and speed limits. Caltrans also maintains the professional services of crash investigation and crash data, as well as knowledgeable 21st-century safety planning. Many useful, but incremental, safety initiatives are constantly being devised and pursued vigorously.

While there is no standard template for organizing a state DOT, many of the common elements impinging on safety tend to apply widely across the states:

- Public-facing safety activities address accepted issues such as alcohol impairment and driver distraction;
- Safety research tends to focus on solving locally recognized problems;
- Safety responsibilities are spread across a number of agency functions; and
- Traffic engineering and ITS have prime responsibility for safety engineering and innovation.

All in all, state DOTs bear significant day-to-day responsibilities in road safety but have a limited role in safety science and innovation. State DOTs carry out the lion's share of traffic engineering and ITS deployment, and therefore have an important influence on road safety. They could have a greater safety impact but are constrained by state government procurement and accountability standards that tend to create a climate of risk aversion. Cyber technology is considered to be a risky use of public funds.

3.5.6 FROM TRAFFIC CONTROLS TO V2X TECHNOLOGY

Traffic engineers are trusted to manage potentially dangerous conflicts between vehicles across the entire road network. Many of these conflicts occur at intersections, interchanges, converging streams of traffic, crossings and in dense urban situations. Traffic engineers have little opportunity to experiment and are discouraged from doing so by tight budgets and having to play catch-up with current technology. America's intersections are not fully covered by conventional traffic signals; and the large number that are equipped are not fully optimized or maintained. Older generation signal controllers operate on fixed phases that need to be retimed periodically; but there are many competing uses for precious highway dollars.

Newer generation adaptive signals automatically adjust phases depending on traffic counts on the various legs of the intersection. The rollout of such adaptive signals is far from complete and remains minimal in many locations. The sensors that initiate adaptation of the signals continue to go through stages of technological change. The older "loop detectors" buried in the road surface tend to be inaccurate and may be unreliable. The newer, more costly, methods using radar and machine vision are more complex and require more powerful lane-specific analytics. Signal controllers may be linked along corridors in order to address higher level goals for traffic throughput, average speed, minimum delay and acceptable queue length. Closed-circuit TV (CCTV) may be installed to monitor wide-area traffic status and incidents. In urban areas manned traffic control centers are used to monitor traffic congestion – both recurring and non-recurring – and to deal with disruptive traffic incidents such as crashes. Control room operators are able to take remote action to relieve choke points via specific intersections and controllers. However, most aspects of real-time traffic control are heuristic and remain relatively unscientific.

While traffic control is in an incomplete and evolutionary state, newer wireless technologies that speak to the system of vehicles and roadways are becoming available. In doing so, they speak to traffic congestion as well as safety. But the next step to highway crash avoidance, using V2X connectivity and potentially automation, is a big one. The impact on state DOTs like Caltrans will be large. And Caltrans already sense that changes will be needed right across the organization, not just within the existing safety groups. To fully deploy the safety paradigm of V2X a chain of actions, involving multiple Caltrans Divisions, is needed. These actions include: (1) organizational change and workforce development, (2) planning and programming, (3) pilot deployments and testing, (4) deployment readiness, (5) external outreach and partnerships and (6) policy development.

It is hard to imagine the size and complexity of the task in mobilizing 50 state DOTs to move beyond the current mode of safety-inclusive activities to the systematic cancellation of highway crashes. These organizations see themselves as having more than enough problems and believe that the substantial cost of V2X deployment would detract from deeply and widely deployed traffic controls. And there is a risk that insufficient numbers of vehicles will be V2X-equipped. On the positive side, they are assisted by the prominent national association AASHTO that can provide technical advice. In addition, regional groupings of state DOTs have knowledgeable associations, such as the Mid America Association of State Transportation Officials (MAASTO), comprising ten midwestern states.

Despite the rightful prominence of safety in all of these organizations' strategies and goals, there are too many degrees of separation between the owners of the highways – the state DOTs – and the vehicle drivers and passengers who are being killed and injured. Beyond road rules, signals, signage, paint on road surfaces and a limited number of variable message signs, what points of contact with road users does a state DOT have at its disposal? V2X has the potential to overcome such current barriers once it is widely deployed in vehicles and in the infrastructure. This difficult chicken-and-egg problem is exacerbated by the substantial costs on both sides of a system that is handicapped by historical fissures: principal among them is the challenge of bridging the commanding private value center of automotive and the irreplaceable public commons of infrastructure.

3.6   ROAD SAFETY METRICS

If our touchstone is harm caused by collisions in the road transportation system, unimpeachable metrics are ground zero for harm reduction. These metrics start with totaling many different forms of mayhem that speak to death, injury, damage, disruption and economic losses. Such forms of harm would apply in certain geographic or political jurisdictions, rural or urban regions, certain terrain or built environment, roadway class, roadway geometry and furniture, traffic volume, safety treatments, speed zone, weather and time of day. They would pertain to different classes, specifications, age and condition of vehicles, unpowered and powered personal machines and pedestrians.

The "host" would be drivers, occupants and other road users of specific age, gender, pre-crash and crash behavior, state of impairment, protective measures, driving experience, driving record, insurance status or socioeconomic group. Great attention is given to the nature of harm to humans, using established scales of injury severity and the final outcomes for humans. Other forms of harm include vehicle loss and repair, lost employment, traffic delay, recovery costs and roadway costs. Harm would be accrued over a specified period, usually in years.

3.6.1   Accrued Retrospective Data and Performance Metrics

Such data is collected after the event and may only be accrued in a meaningful way over a substantial period such as 12 months. Use of the data and the metrics may involve a wide range of agents and a range of purposes. The most noble of these purposes would be development of countermeasures to individual forms of harm, caused by certain crash types. This would probably require referencing total harm of a particular kind to the host's exposure, either broadly or to the relevant traffic situation. Other purposes would include jurisdictional reporting on crash rates, for consideration in road safety planning and assessment. Broad metrics used by safety agencies such as NHTSA include annual fatalities per 100 million miles (VMT) and annual fatalities per 100,000 population.

In 2018 NHTSA published a manual (12) containing selection criteria for road safety performance metrics. At a broad level, three-year or five-year moving averages of fatalities per 100 million Vehicle Miles of Travel (VMT) are recommended. VMT should be extracted from FHWA's Annual Highway Statistics. In specific examples, metrics could be assessed for a wide range of niche purposes; for example, it is possible to access measures for fatalities of unhelmeted motorcycle riders, or drivers aged under 20 years, and many others.

The General Motors-based traffic safety scientist Leonard Evans was a strong advocate for "action based on understanding" (13) and sought metrics and statistical methods that provided insights into the effects of policy environments and specific national and regional circumstances. He also used the crash data to address many questions of crash risk related to driver behavior, impairment, restraints, on-board safety devices and the roadway. He pioneered innovative means of statistical inference based on the relative probabilities of harm subject to specific variables, such as impaired versus unimpaired drivers. Analysts like Evans provided a rich field for countermeasure development by others. This led to a modest industry built upon safety features. Much of this work is done by automotive suppliers, including the large Tier 1 suppliers like Bosch and more specialist developers like Hella.

3.6.2 THE CONCEPT OF DRIVING RISK

As the term became more popular in the late 20th century, the notion of "risky driving" materialized. At the present time, NHTSA recognizes the following forms of risky driving: drunk driving, drug-impaired driving, distracted driving, speeding and drowsy driving. In a similar vein, Australia's 2022–2030 national road safety strategy recognizes forms of "risky road use" above and beyond illegal actions, including distracted, fatigued and inattentive driving. Current national strategies try to minimize or eliminate these forms of behavior. However, it should be apparent that the term risk is fluid when applied in road safety.

Road safety has a constant need to prioritize the application of limited resources, and it makes sense to address the greatest risks sooner. Quantitative methods are needed. By the time of the 1990s, mature databases of fatality and serious injury circumstances and outcomes had been established in many countries. In order to quantify the harm caused by specific types of motor vehicle crashes, analysts in Australia and Sweden adopted Rumar's 1999 "ice cube" analogy (14) of three axes of quantification: exposure to a particular type of crash (usually on the basis of distance traveled), risk of a crash of that type occurring and consequences of that type of crash (principally in serious injury or death). The product of these three metrics (exposure *times* risk *times* consequence) would then represent total harm caused by this particular type of crash, over a specified period of time. The crash type would be defined by the host (the driver, vehicle occupants and other road users) the vehicle and the roadway class and environment.

Many formulations of risk have been used in road safety, although the basic concept that has been used consistently is: the probability of a harmful event occurring per unit exposure to that type of event. The idea is that the harmful event (the first collision) is not certain, but its likelihood may be calculated based on historical crash data. Once the event has occurred, the probabilities of various severities of resulting injury or death (the second collision) may be used to calculate the crash consequences in aggregate. Such use of the term "risk" is somewhat informal, and there are many other uses and interpretations in road safety.

Given that the consequence of fatality is usually of greatest concern, the remaining two axes – exposure and risk – have endured as the indispensable metrics of road safety analytics. This has not been without its problems because crash databases tend not to contain exposure information. Vehicle Miles of Travel data tends to be collected by traffic departments who are interested in traffic management and planning. It may not be available in categories that match the needs of safety analysis and more localized decision-making. Many uses of exposure data were discussed by the Dutch Institute for Road Safety Research (SWOV) (15). In some cases where crashes tend to involve multiple vehicles, traffic flow provides a more appropriate exposure measure than miles traveled. And hours of use may be better than distance covered for certain purposes: for example, in comparing transportation modes that differ greatly in velocity, such as road and air travel. In a further twist on the use of countermeasures, Borkenstein's well-respected 1964 study (16) of driving risk as a function of Blood Alcohol Concentration (BAC) the author extracted information on BAC levels of a large group of drivers involved in crashes and similar information on drivers who did not crash. He assigned a value of 1 to the relative risk of zero BAC and derived an exponential-looking curve that showed a doubling of relative risk when BAC approached a value of 2. When the relative risk is greater than 1, it is termed "over-involvement", a term that has since permeated road safety very widely.

In recent decades, risk assessment has grown in importance in many fields, including engineering, finance, insurance, the environment and education. Risk mitigation has become an essential part of modern governance and project management. The use of the term in road safety has remained straightforward and non-technical. At its root, risk is as much a philosophical concept as an engineering management tool and is highly relevant to the philosophy of technology. According to the Stanford Encyclopedia of Philosophy (17), risk combines several strands of ethical theory and is highly adaptable. Certainly the safety touchstone of utilitarianism allows us to use probability-weighted predictions of outcomes. But moral considerations must also be included; it may be appropriate to distinguish between risk-taking and risk-imposition, who contributes to the risk, and with what intentions, and how the risk and benefits are distributed. Individuals have a certain right not to be exposed to risk of negative impact, although the practicalities of life then raise the need for exemptions. In response to this issue, reciprocal exchanges of risks and benefits could be considered, such as: "if A is allowed to drive a car, exposing B to certain risks, then in exchange B is allowed to drive a car, exposing A to the corresponding risks" (17). It is fair to say that such thinking has not received serious attention in road safety; future changes in the transportation system, such as automated vehicles (AVs), will require a more nuanced view of risk and morality.

3.6.3 INTERNATIONAL COMMUNITY OF ROAD SAFETY ANALYTICS

Road safety is a global problem, and safety professionals in many countries have exchanged philosophies and methods. Techniques in the collection and analysis of crash and injury data have been shared for many years. Increasing commonality has enabled comparisons to be made. But philosophies and motivations can differ. The United Nations declared 2011–2020 as a decade of action for road safety, with a target of reducing deaths and serious injuries by 50%. A second decade of action was declared by the UN General Assembly for 2021–2030 with the same goal of reducing deaths and injuries.

This mission of significant harm reduction is treated as a public health issue, and the World Health Organization (WHO) has developed a strategy to address the UN goal. Safety data is being used for development of strategy, identification of solutions, for advocacy and monitoring, analysis and evaluation. In the context of such a campaign, safety data categories are expanded to act more like Key Performance Indicators (KPIs) as used in modern business management practices. Key safety data includes exposure data (traffic volume, population data), final outcome data (deaths and injuries) and intermediate outcome indicators (average speeds, protective equipment fitment and use, level of drinking and driving, network and vehicle safety quality). Regions and countries are encouraged to implement their own version of the strategy. Perhaps due to a political sensitivity to highlighting the performance of individual countries, the UN places more emphasis on progress in aggregate.

However, there have been many comparisons between rich developed countries. Safety advocates from one country may like to use crash and injury rates from other countries to make a point about the need for more research funding, road investment and the like. In 2013, the developed-country economic organization known as the Organisation for Economic Cooperation and Development (OECD) published a list of key safety statistics pertaining to all of its 37 member countries (18). Table 3.1 shows the numbers for several selected countries, including the United States. This sort of data tends to support the idea that a certain standard of road safety performance is endemic to high-income democracies (Australia, Japan, Sweden, United Kingdom and United States) and a lower level of safety performance applies in less democratic countries (Malaysia and Russia). Such comparisons often consider deaths per unit distance traveled – rather than per unit population – because some countries like the United States do a lot of driving.

The metric of deaths per 10 billion vehicle-km brings out the idea that each country's road and traffic system has a certain engineering and behavioral quality in relation to safety. Then the total safety consequences depend on how much that system is used – the number of drivers and the amount of driving. National safety quality does vary between rich democracies, and those that strive the hardest, like Sweden – the home of Vision Zero – do twice as well. Those democracies that drive the most, like the United States, are still two or three times better than the illiberal countries like Russia. Of course, such metrics should be compared with caution

TABLE 3.1
Road safety metrics in selected OECD countries (2011 data)

Country	Deaths per million pop.	Deaths per 10 billion veh-km	Deaths per 100,000 registered veh	Registered vehicles per 1,000 pop
Australia	51	50	7	751
Japan	40	69	6	657
Malaysia	231	122	29	792
Russia	124	201	36	353
Sweden	27	34	5	597
United Kingdom	28	35	5	551
United States	103	68	12	852

TABLE 3.2
Road fatalities per 100 million km of travel (19)

	2013	2014	2015	2016	2017	2018	2019	2020	2021
Australia	5.3	4.7	4.8	5.1	4.7	4.4	4.6	4.5	4.6
Japan	7.1	6.7	6.8	6.4	6.0	5.6	5.2	5.0	4.9
Norway	4.3	3.4	2.6	3.0	2.3	2.3	2.3	2.1	1.8
Sweden	3.3	3.4	3.2	3.3	3.0	3.8	2.6	2.6	2.6
United Kingdom	3.6	3.7	3.5	3.4	3.4	3.3	3.4	3.5	3.4
United States	6.8	6.7	7.1	7.3	7.2	7.0	6.8	8.4	8.3

because there are many issues with the collection and reporting of crash and injury data with a worldwide scope.

The OECD is a reputable international observatory of road safety statistics. They do not seek publicity and often take on very unpopular topics that nevertheless benefit from unbiased quantification and analysis. The key safety metrics in Table 3.2 show annual OECD figures through the latest decade available, for six selected countries. The measure is our chosen indicator, but in metric units: deaths per billion vehicle-km. Rather than focusing on comparisons between countries, we would want to see that measure, and others, reducing over time and we do observe improvement for some countries, but not all. The figures do serve to illustrate the difficulty of setting aggressive decadal improvement targets either at the national or aggregate level.

3.7   DRIVER EDUCATION AND COMPLIANCE

3.7.1   ROAD SAFETY CAMPAIGNS

The Australian state of Victoria has a history of purposeful road safety interventions, including media campaigns targeting certain high-risk behaviors, such as speeding and drink-driving. Victoria established the Transport Accident Commission (TAC) in 1987. It is a statutory insurer of third-party liability. A portion of each vehicle's registration charge is reserved to TAC who provide support to the victims of road crashes. TAC also produces graphic TV commercials that seek to dramatize chains of events leading to crashes and highlight egregious behavior and resulting remorse. In its heyday, TAC coordinated with police and road patrols to associate the violent consequences of bad driver behavior with on-road enforcement and severe penalties. It is believed that the driving public are more accepting of large fines if they understand the reasoning behind them. In the latter part of the 20th century, TAC succeeded in planting safety sentiment in the vernacular of Victorians with tag lines like "If you drink then drive, you're a bloody idiot" and "Don't fool yourself, speed kills".

In a similar vein, NHTSA has maintained a long-time campaign, called "Click It or Ticket" to encourage the wearing of seat belts. While it was up to individual states to activate such campaigns, some states carried out simultaneous launches of this campaign with an upgrade of seat belt law enforcement from secondary to primary enforcement. Today, all states carry out specific campaigns that fall under a current rubric of promoting a safety culture. For example, the California Office of Traffic Safety (OTS) says:

Let's be the best versions of ourselves on the road, especially if we are in the car. Bicyclists and pedestrians need drivers to be safe to keep them safe. And drivers need those around them to exercise care. Share the road. Share the responsibility. Let's look out for one another.

Important actions promoted as sharing the road include:

- Following the speed limit;
- Slowing down at intersections;
- Being prepared to stop and let pedestrians cross;
- Taking care in less-than-ideal driving conditions such as in the dark, foggy or rainy weather; and
- Respecting bicyclists and giving them space when passing.

Campaigns carried out by OTS, and typical of such campaigns all round the world, cover drink driving, distraction from personal devices, proper use of child seats, slowing down, wrong-way driving and teen driving. Such campaigns are prompted by surveillance of crash data and reacting to disturbing trends. Safety officials are expected to oversee small but steady improvements over time. Certain types of crashes, defined by demographics or locations, may buck the trend. For example, California's pedestrian and cycle fatalities increased by 50–60% over a decade and prompted initiatives by OTS.

In April 2023, OTS launched a campaign targeting distracted driving, called "Get Off Your Apps". This launch was timed to coincide with National Distracted Driving Awareness Month, and NHTSA ran a national media campaign, paired with a law enforcement crackdown, called U Drive. U Text. U Pay. The deliberate combination of increased awareness and increased enforcement has evolved from successful safety campaigns all over the world. But NHTSA provided a sobering message at the launch of this particular campaign:

Traffic fatalities and the fatality rate declined consistently for 30 years, but progress has stalled over the last decade and went in the wrong direction in 2020 and 2021. Today, NHTSA is releasing our updated 2021 traffic fatality statistics.

In 2021, fatalities increased by 10%, with 42,939 lives lost on our nation's roads. This is the highest number since 2005 and the highest percentage increase since NHTSA's Fatality Analysis Reporting System began collecting data in 1975.

The fatality rate per 100 million vehicle miles traveled increased to 1.37, a 2.2% increase from what we saw in 2020. And fatalities increased in 43 states, the District of Columbia, and Puerto Rico.

The risky driving behaviors that increased in 2020 during the pandemic continued into 2021, with speeding, impaired, unbelted, and distracted fatalities all on the rise.

Vulnerable road users bear a heavy burden from distracted and other risky driving behaviors. Pedestrian fatalities increased 13% in 2021, and cyclist fatalities rose by 2%.

3.7.2 GLOBAL APPETITE FOR SAFETY CAMPAIGNS

At the global level, the United Nations and the WHO is promoting a second decade of action for road safety 2021–2030. The stated goal is to reduce road deaths and injuries by 50%, and the strategy includes recommendations for new policies and rules to be adopted by nations and responsible jurisdictions regarding:

- Better post-crash care;
- Enhanced driver qualification frameworks;
- Expanded legislation regarding seatbelt use, child restraints and alcohol and drug driving; and
- Accelerated adoption of standards mandating inclusion of new vehicle safety technologies.

Such global campaigns necessarily cover a huge range of societies, from developed, rich countries to various levels of developing countries. Suffice it to say, these efforts are mostly based on a paradigm of drivers causing crashes and therefore drivers needing to change their behavior. Many of these campaigns are developed and operated by the same government agencies that are responsible for the design, operation and maintenance of road systems and for the enforcement of road rules such as speeding. Continuous monitoring of trends in crash statistics by these organizations is part of their responsibility in road safety.

If we step back and think about the main components of road safety – vehicles, drivers and roads – we see that most road safety campaigns are directed at faulty drivers. It is perhaps ironic that the government agencies who offer advice to drivers are not subject to similarly independent scrutiny. The other party – the collective of vehicle manufacturers – is of course subject to intense regulatory scrutiny and have been the agent of most of the road safety improvements over the past 50 or 60 years.

While these campaigns have access to good data on crashes and their consequences, they have become marketing exercises over time and have commoditized safety. The topics covered in campaigns include a mix of injury prevention initiatives (such as seat belts and child car seats) and crash avoidance issues, especially impaired and distracted driving. They also conflate messaging for greater awareness, especially for pedestrians and cyclists, with actual countermeasures, such as technologies for emergency braking. Overall success with national and global safety campaigns has been limited at best relative to their targets.

3.7.3 Morality of Safe Driving

Public messaging has also increasingly appealed to drivers' moral responsibility not to expose other road users to risk and harm. Risky driving may be portrayed as a black mark against good citizenship, but is hardly top of mind for individuals, given the challenges they face to make their way in a competitive world. That world is increasingly stratified and quantified and each person is likely to be rated for intelligence, knowledge, sporting prowess and life skills such as financial management. Safe driving is not on an individual's critical path for leading a successful, happy life. Even for a thoughtful individual, it is not a virtue they work hard to acquire.

From a philosophical point of view, road safety has many ethical attributes. By far the most dominant is utilitarianism, our yardstick being Greatest Happiness for the Greatest Number (GHGN), as famously advocated by John Stuart Mill (9) in the mid-19th century. As our world has seemed to downplay happiness as a meaningful goal, and we do not wish to be accused of hedonism, many may prefer the economist's term "utility", rather than "happiness". The utilitarian's measure also includes the absence of pain, and this is easily recognized in the reduction of road crashes and injuries, and their broader consequences. So we are able to recognize a certain morality in the utility and life-importance of driving, along with our ability to reduce crashes and injuries. There is also morality in the continuous improvement in the science and application of safety.

However, traffic rules, and messaging for safe driving, remain quite remote from harm reduction. The messaging worldwide is remarkably similar, and there is no doubt that impaired driving and high-speed driving increase the risks of collisions. And the continuing pressure for wearing restraints such as seat belts is absolutely justified. But these entreaties do not rise to the level of morality and do not teach us much about the virtues of driving. What does it mean to drive with a good will? Driving is almost always carried out in the service of other front-line life activities – such as employment or education – and therefore goes wherever those pursuits take it. Driving with a good will requires knowledge and skills specific to collision avoidance, and this varies from place to place and from time to time.

We have come far from the mid-century days of motor vehicle deaths and injuries as an affront to the medical profession and in turn the automotive industry. These and other professional sectors could not condone the "epidemic" of automotive injuries. By this authorities meant that violent injuries were increasing over time, perhaps out of control. Many professionals were well accustomed to working with ethical guidelines and high standards of practice, as appropriate to their chosen sphere of vocational commitment. The fact that horrific injuries were no longer rare events was vocationally intolerable, and certain professionals worked very hard to change things for the better.

Road safety has now reached a very different place. Despite some loose statements by activists, we are no longer in an epidemic. Gross statistics are not increasing, but neither are they decreasing as we would like. The onus has shifted from the front-line road safety professions – especially medicine and engineering – to the driving public. And that is where road safety campaigns are directed. On the face of it, this seems reasonable, but focusing on drivers causing the vast majority of crashes may be missing the point. These days, many would say that each crash is "contributed to" but not caused. We have created a transportation system of many elements and narratives, and unstoppable momentum. Its complexity demands many levels of abstraction and a system understanding. For example, the transportation system might be described as: "the combination of elements and their interactions, which produce the demand for travel within a given area and the supply of transportation services to satisfy this demand". How are we to square such a bloodless definition with the troubling number of fatalities and serious injuries endured every year? We are not dealing with a closed, professionally run system that is amenable to moral arguments.

3.8   REPAIRING THE DAMAGE

3.8.1   PROPERTY DAMAGE ONLY (PDO)

We have focused on motor vehicle and roadway mayhem in terms of death and injury, but other forms of damage are also significant. Crash statistics, and ana-lysis thereof, are most relevant to deaths and injuries, because investigation of crashes tends to concentrate on the human consequences. Another category of consequences – usually termed Property Damage Only (PDO) – are reported, but with less detail and lower reliability. In 2019, NHTSA reported 4,374,000 PDO crashes involving 6,957,000 vehicles (20). The total park of registered vehicles was 253,000,000. A more detailed NHTSA analysis of PDO crashes from 2010 found 18,508,632 vehicles damaged in crashes, and most of these were not reported to police.

The sheer number of damaged vehicles adds up to a very high proportion of the total economic cost of all crashes: 31%. This PDO cost is the largest component of crash cost, with congestion costs at 12% and medical costs at 10%. It is remark-able to consider that some 7% of all U.S. vehicles registered are damaged in crashes each year. While there have been few attempts to tie vehicle damage to roadway characteristics, NHTSA shows that the largest cost component is incurred at urban intersections.

3.8.2 WHO PAYS FOR CRASH COSTS?

NHTSA's statement (20) concerning the burden of crash costs is well worth quoting:

> Private insurers pay approximately 54 percent of all costs. Individual crash victims pay approximately 23 percent while third parties such as uninvolved motorists delayed in traffic, charities and health care providers pay about 16 percent. Therefore, those not directly involved in crashes pay for over three-quarters of all crash costs, primarily through insurance premiums, taxes and congestion related costs such as travel delay, excess fuel consumption and increased environmental impacts. In 2010 these costs, borne by society rather than by crash victims, totaled over $187 billion.

Therefore, the majority of crash costs are spread across a vast number of insured automobile owners, while the driver actually involved in the crash pays only a small amount. And what of the government agencies responsible for the road network that hosts so much destruction? They are barely mentioned, either as incurring losses or contributing to costs.

As we ponder the great question of who is responsible for the road transportation system, we may care to notice that the NHTSA report tells us that vehicle insurers collectively paid out $131 billion in crash costs in 2010. In the same year, federal budget sources tell us that the federal highway bill authorized $40 billion for highway programs and $335 million for NHTSA. If we also take into account the $139 billion spent by state governments, as included in a 2018 Forbes article (21), crash costs are equivalent to a massive 70% plus of all expenditures on the nation's roads. In other words, the high level of destruction taking place on the system goes a long way to cancelling out – at least in a moral sense – the constructive national effort to improve the network and counter the crash problem. We need to realize that, each year, all taxpayers chip in for the $180 billion cost of roads. On top of that, all drivers who insure their cars pay into the $130 billion account of crash money. This account, or set of accounts, is privately held and only drivers involved in crashes get to make withdrawals.

Again, the Australian State of Victoria provides an interesting road safety twist. A portion of each vehicle owner's annual vehicle registration fee is earmarked for the Transport Accident Commission (TAC), a state government body whose main job is to compensate crash victims for personal injuries. So the system of payouts to folks injured in crashes is run by the government, rather than by private sector insurers. Recent figures from TAC (22) and the state road agency show that about A$3 billion is spent on Victorian roads each year, while TAC pays out about A$1.5 billion to crash victims. In other words, crash injury costs are equivalent to about 50% of road expenditures. In the United States, all payouts to crashed drivers – whether injury-related or vehicle-related – come from the private insurance industry. In 2022, that market was estimated to be $328 billion (23), a significant adjunct to the total value of the U.S. auto industry ($1.5 trillion in 2021).

We see a general principle that the entire class of taxpayers pay for roads, while the entire class of drivers pay, quite separately and fairly completely, for crashes. In the United States, road crashes place a 20% surcharge on the entire automotive industry. And that surcharge is paid by the auto industry's customers, as an unquestioning class, including the great un-crashed. This expensive form of public–private mayhem involves only the crashed few, weaving unseen among the huge mass of vehicle owners.

The toll extracted from a population by road crashes is frequently publicized in terms of the human consequences, including life and quality of life, but the sheer scale of the dollar cost is not widely appreciated. Whose job is it to take account of this situation, which takes on a moral, as well as economic, dimension? It is obvious that the road transportation system is not a closed, professionally managed system. If it was, the disproportionate and hidden crash cost would be unethical and probably negligent. Suffice it to say that road systems are basically run by state governments, and the ambulances and tow trucks operate in plain sight, under their purview. The insurance industry also acts as a pass-through mechanism for a vast dollar amount of crash damage to people and property.

3.8.3 THE GREAT UN-CRASHED

We should question the ethics of the crash cost equation. Many drivers – we might call them the "great un-crashed" – who have never had a crash and therefore have not made a withdrawal from the crash bank, may not understand that all of their fees go directly to the few that do crash. And also that the only money the crashed few receive is from the crash bank. There is no other cash contribution from the U.S. government. The un-crashed majority has very little say in this.

We may conclude that the philosophy of road safety, despite showing great virtue on the surface, does not contain even the beginnings of an overt social contract. Such a contract could be expected to consider rights and responsibilities of users, rights and responsibilities of the state, fairness, open information and well-founded belief and pride in safety improvement. Everyone should be able to see that the parties to the road transportation system are all doing the right thing.

3.8.4 BIG SAFETY

Is there such a thing as the "safety-industrial complex"? Has road safety as a total enterprise risen to the level of an economy, with R&D, suppliers, markets and the like? Market research companies suggest a current global market of several billion dollars, through the supply of intelligent safety systems to public road and trans-portation agencies. A somewhat separate, and larger, market has existed for some time on the automotive side, supplying vehicle crashworthiness technology and crash avoidance technology. Research companies suggest a current global market value of $20 billion for one growth element of this technology: ADAS.

More traditional automotive safety products such as airbags already command similar markets. The Swedish company Autoliv, which started out as a seat belt manufacturer, is a well-established $8 billion company that acts as a Tier 1 automotive supplier specializing in safety equipment. Overall, the worldwide Tier 1 market is estimated to be in the vicinity of $500 billion, so the safety-related component still remains modest.

It is clear that any suggestion of a safety-industrial complex would point more to the insurance industry. They enable the massive cost of road vehicle crashes to be covered by shielding the small percentage who do crash from the costs they actually incur.

Big changes are afoot in these markets, including the rise of battery electric vehicles (BEVs) and the development of driverless car technologies. There is still great expectation around AVs and global market estimates rise to several trillion dollars. Because the AV is promoted extensively as an unparalleled safety technology, the safety-industrial complex is expected to change rapidly and expand. The development of the ADAS market is a forerunner of this development. Advanced computer chips, such as the successful products of NVIDIA, are already available to make this happen.

These automotive-based markets far exceed road safety markets in the infrastructure world. Infrastructure markets are served by a different set of companies: those in the business of traffic control and ITS. Deployment of ITS systems is limited by public sector procurement issues, and this limits the growth of the ITS industry. ITS was conceived in the late 20th century to maximize the productivity of existing infrastructure and to hasten the connectivity between vehicle and infrastructure (particularly the intersection) and from vehicle to vehicle – to vehicle. In other words, to create a workable system where the safe control of one vehicle could be informed by the movements of other vehicles and by more responsive traffic controls.

But serious 21st-century efforts to bring forth the coordinated technologies needed in vehicles, and also at intersections, have so far failed to leap over the public–private divide. Not the least of the barriers encountered was lack of common understanding between the makers of the cars and the keepers of the intersections. Would one side or the other be left with cyber egg on their face through lack of standardization and a stable plan?

Commercialization of safety sub-systems for automobiles has certainly increased in the 21st century but is not out of context with other R&D developments. The market has high standards and is largely conducted through, or in partnership with, the existing tiered network of suppliers. Automakers have consistently maintained that safety is "pre-competitive". That is, manufacturers do not try to compete on the perceived safety of their makes and models. At the consumer end, manufacturers do exert their individuality through the active safety systems – above and beyond the safety elements required in federal motor vehicle standards – that they offer their customers. Systems such as side object detection, whereby a driver is alerted to the presence of a vehicle occupying their blind-spot, may be offered as options in the new vehicle and may be packaged with other similar systems. Manufacturers offer distinctive sets of such options, often given names like "safety package" etc.

By far the largest global market related to road safety is auto insurance, estimated to have a total value of $650 billion in 2021. This looms quite large relative to the total global auto market, now approaching $3.0 trillion. Otherwise, there is no evidence that road safety is being oversold, purely to create a large addressable market that could be manipulated by public relations and advertising.

3.8.5 INSURANCE AND LIABILITY

In 2019, Kyle Logue of the Cato Institute proposed a radical new combined system for auto insurance and liability (24). In selling a vehicle, an automaker would assume responsibility for all crash costs incurred in and around that vehicle after it enters service on the road system. This was termed "automaker enterprise liability". Part of Logue's premise is that the current patchwork of auto insurance and product liability does not provide any clear safety motivation to drivers. Auto insurance, so carefully nuanced with all its vicissitudes inside the company, is not well oriented to reduce driver's crash risks because the company and its customer have totally different realities with regard to crash risk. And while automakers are subject to auto liability laws, the nature of these laws and the actual meaning of negligence is not generally apparent to vehicle owners and the driving public.

The asymmetric state of safety information is well expressed by Logue as he sets out the case for change:

Auto insurers are, unlike most drivers, extremely well-informed about the intricacies of accident law. They employ teams of lawyers whose job is to understand how driver liability laws in each state affect the liability risks of their customers. Indeed, their profitability and their survival as going concerns depend on this expert understanding of the auto liability laws of all sorts. In addition, auto insurers have unparalleled access to enormous amounts of detailed information regarding the crash-risk characteristics of millions of drivers and automobiles. This is the result of decades of experience providing auto insurance coverage to hundreds of millions of drivers and vehicles, which in turn means pricing millions of auto insurance policies and adjusting millions of auto-crash claims over the years. No other institution or organization has the same amount of driver-specific and automobile-specific data as the auto insurance industry.

According to Forbes (25), 79% of U.S. drivers carry comprehensive car insurance. The insurance products they buy are the sum of the internal efforts of one of the most-researched industries on earth. Unfortunately that research is internal and closely held. The structures of the policies offered are complex and include elements of annual fee, discounts, deductibles and co-pays. Aspects of vehicle usage also play into the amount the consumer actually pays. On the side of the company, customer data such as policy history, claims history and location are just the start, with personal information also coming into play. This will include credit history, educational attainment, gender and marital status. Much of this is not apparent to the customer; on the other hand, many relevant aspects of the customer's lifestyle and life performance are kept close to the customer's chest. The negotiation, such as it is, is carried out with very few cards on the table. It is not surprising that car insurance is not unequivocally known as a force for good in road safety.

3.8.6 EFFECT OF INSURANCE ON SAFETY

The commercial application of insurance has existed for centuries. Certain tendencies have been questioned on ethical, as well as economic, grounds. In addition to the problem of asymmetric information – whereby the two sides are required to make a contract while effectively sitting in different rooms – there is also evidence of moral hazard. This refers to some customers taking greater risks when they feel protected from adverse personal consequences. Another endemic flaw is adverse selection, where the company attempts to penalize a customer based on certain of their characteristics. In the case of auto insurance, moral hazard simply means that the insured may drive in a more risky manner – say drive a little faster – than they otherwise would. This sort of thing is difficult to prove, but studies have indicated that increased insurance correlates with increased crashes. In other words, an adverse safety trend.

Automotive insurance claims rose to a total of $197 billion in 2021 (23). This level of cost underlines the important place of auto insurance in the road safety economy, even though its effect on road safety remains obscure. That obscurity may be changing, with some vehicle owners opting to connect a vehicle sensor that provides auto insurers with meaningful data about the way a vehicle is actually driven. This approach is called usage-based insurance (UBI) in the industry and is enabled by connected vehicle technology. It means that insurance bills could track downward with less risky driving, providing a clear financial incentive for driving safety improvement – a positive safety trend.

3.8.7 CHANGING LANDSCAPE OF AUTO INSURANCE

The prospect of big changes in an industry can often illuminate the way it actually works now. Extensive R&D in driverless cars over more than 10 years has prompted examination of many auto safety issues that were lying dormant. The insurance industry has been forced to consider how driving risk and liability may transform and the possibility that product liability may take over at least some of the space currently occupied by insurance. Because some automakers have already made statements about accepting the crash risks of their AV products, attention has been devoted to the question of who is liable for what. It may be that automakers will willingly accept more product liability, rather than enter a protracted debate about AV fault and risk. That debate would potentially delay the entire rollout of a new product where a commercial payback is eminently overdue.

From the earliest days of motor cars supplanting horse-drawn conveyances, there was a need to attach blame when an "accident" occurred. Police learned to quickly identify a culprit and begin the process of retribution. It became necessary for vehicle owners to be insured, and the rest is history. Insurance was relatively simple to understand and potentially covered owners for a multitude of sins. On the other hand, it was always going to be difficult to pin a crash on a vehicle defect. This required expertise at all levels of the process of justice; it was going to be tricky and probably more expensive. So the actions of drivers became the first port of call for crash investigators.

3.8.8 PRODUCT LIABILITY

In the latter part of the 20th century, product liability and associated legal structures became an American pre-occupation, under laws that emerged state by state. Damages may be sought from the manufacturer for personal injury or property damage, and claims may be framed as product liability, negligence or breach of warranty. Product liability litigation based on deaths and catastrophic injuries have resulted in large awards against U.S. automakers but not necessarily in a way that will benefit road safety. The size of an award may not provide a clear safety signal. For example, defendants that appear to be uncaring can lead to very high compensation to plaintiffs, but this would not affect their motivation to drive safely. Product liability could simply lead to corporate fear, instincts toward self-protection and inaction.

In a more measured way, automakers are also subject to vehicle recalls, sometimes based on repeated complaints channeled via NHTSA. These recalls may be required by NHTSA or initiated voluntarily by the manufacturer. Auto suppliers frequently become involved in such recalls. Automakers have become pro-active and risk-averse when it comes to defects, preferring to work in a professional way through NHTSA rather than taking their chances in court. While integrated data on automaker liability is not available, NHTSA's efforts in aggregating complaints, investigating safety design oversights and managing vehicle recalls have become a major part of their work. The rectification of vehicle safety defects through all of these channels has had a positive influence on road safety, both in an objective sense in marginally reducing crash risk and in increasing consumer confidence.

3.9 THE ADVENT OF INTELLIGENT TRANSPORTATION SYSTEMS (ITS)

As one of its founders, Joseph Sussman of MIT saw ITS as part of the "Complex, Large-Scale, Integrated, Open System" of transportation (26). By 1991, the U.S. Interstate System would be virtually complete and a new vision for the transportation system would be needed. The 1991 re-authorization of the federal-aid highway bill set the scene for ITS as we know it today. As a result Sussman was responsible for drafting a strategic plan for the Integrated Vehicle Highway System (IVHS) of the future. ITS was sought as an alternative to more and more physical construction, in order to address rising levels of traffic congestion and to improve safety. The strategy also recognized the rising importance of environmental issues and the need for increased productivity to maintain the global competitiveness of the U.S. economy. In other words, safety was an integral component of ITS, but there was also a lot more to it.

3.9.1 DEMO '97

A new advisory committee called IVHS America was formed within the U.S. DOT in 1990, and it produced a 20-year blueprint for a substantial ITS R&D program.

Part of that program was the Automated Highway System (AHS) that envisaged close platooning of vehicles in special highway lanes. In August 1997, the National AHS Consortium brought this to reality on lanes of the Interstate Highway 15 in San Diego, California. A lasting impression was created by eight close-coupled passenger cars traveling at highway speeds – a fully cooperative platoon. This event was called Demo '97.

The FHWA described the AHS as follows (27): "Highways of the future may feature relaxed drivers talking on the phone, faxing documents, or reading a novel while an automated highway system controls the vehicle's steering, braking and throttling and allows for 'hands-off, feet-off' driving". Note that the onus was on the highway – the infrastructure – to control the vehicles in a safe manner. The heart of the test was a 12-km stretch of the I-15 High Occupancy Vehicle (HOV) lanes north of San Diego. A total of 93,000 magnets were longitudinally installed in the roadway to provide contactless guidance to the platooning vehicles. The AHS consortium brought strong and diverse technological capabilities, combining automakers, aeronautics, defense, suppliers to these industries, metropolitan transit agencies and academia. The work was further described as follows:

> Automated lane keeping, for example, will be demonstrated in a variety of ways: one scenario follows magnets imbedded in the roadway, another tracks a radar reflective stripe, several others use a variety of vision-based systems. And the list goes on and on: laser-based systems, radar applications in obstacle detection and lateral control, advanced communications concepts, ...

The FHWA recognized that full AHS may well be years away, but certain near-term applications were foreseen. These included adaptive cruise control, detection and avoidance of other vehicles and obstacles in the roadway and lane-keeping. The more specific of these applications – adaptive cruise control and lane-keeping – have since entered automotive markets reasonably widely as available options. And many of the sensing technologies have found their way into a wide range of ADAS and are contributing to the development of Automated Driving Systems (ADSs).

3.9.2 COLLISION AVOIDANCE METRICS PARTNERSHIP (CAMP) LLC

Based in Farmington Hills, Michigan, CAMP is an affiliated R&D arm of several automakers, led by the Ford Motor Company and General Motors. Since its establishment in 1996, it has largely operated under research agreements with NHTSA – and later FHWA – to create meaningful technical platforms for Vehicle-to-Vehicle (V2V) and Vehicle-to-Infrastructure (V2I) applications. This work has included substantial contributions in underlying systems engineering and security. As more and more cutting-edge vehicle safety R&D has been directed to bigger challenges – toward automated driving machines– CAMP has provided an important technical bridge between safety advancements in the worlds of automotive and traffic control. For example, current work in their V2I consortium is making an important intersection safety application – Red Light Violation Warning – more accurate in terms of the timing of the message and the split-second locations of the critical vehicle actors.

3.9.3   ITS AMERICA

IVHS America became ITS America, a non-profit independent entity based on an institutional membership model linking companies, government agencies and academic institutions. ITS America provides convening and technical services across the enterprise of ITS and in support of transportation system development. Most importantly, it fosters an associative approach to technology deployment by government agencies, who are by nature risk-averse. ITS America also provides wide coverage through its network of state chapters which bring in many more companies, agencies and universities who may tend to operate regionally rather than nationwide. ITS America also works with the U.S. DOT's Joint Program Office which is active in the deployment of ITS across all of the department's modal administrations.

The important role of ITS America in bringing together all of the key elements of the ground transportation system has endured and prospered. Like road safety in general, there has been an over-reliance on vehicle-based technologies and government agencies have shown a limited appetite for deployment in the infrastructure. Nevertheless, the ITS industry in service of traffic engineering has expanded and we are seeing more advanced technology such as LIDAR sensing being transferred from the automotive sector to the roads sector. This cross-fertilization set the scene for more ambitious traffic systems thinking in the new century, including V2V and V2I. For example, in the case of V2X, the same companies supply the radios to the automakers and to the traffic engineers.

3.9.4   ITS JOINT PROGRAM OFFICE (JPO)

The JPO was created in the 1991 Intermodal Surface Transportation Efficiency Act and continues to play an important role to this day. The Act was signed by President George H.W. Bush and was welcomed as the first transportation bill of the post-interstate era. Clearly the framers of the vision that covered the IVHS and the AHS saw ITS as an inclusive rubric. As ITS programs were created under the JPO, it is notable that the definition of ITS was never strongly articulated. In the new era beyond concrete, technologies were going to be fast-moving. In the current era the JPO says that "ITS is an operational system of various technologies that, when combined and managed, improve the operating capabilities of the overall system".

Relative to Joseph Sussman's idea of transportation as on open system, ITS is not *the system* but it is the glue that brings the pieces together for the purpose of improvement. One of the first major tasks of the JPO was the National ITS Architecture that eventually created a body of standards with three layers:

*   Communications;
*   Transportation solutions including functionality and data requirements; and
*   Institutions and policies for implementation.

Subject to such standards, ITS rolled out many branded initiatives of local or regional application and providing bounded functionality. These initiatives addressed aspects of tolling, traveler information, navigation and commercial vehicle operations.

The 1990s saw the launch of FAST-TRAK, TravTek, Pathfinder, Guidestar, Advantage I-75, INFORM, Smart Bus and EZ-Pass, among others. Although a number of these programs tended to fall by the wayside over time, many seeds were planted – technologies came to wider attention and relationships with users were explored.

3.9.5 DEDICATED SHORT RANGE COMMUNICATION (DSRC)

An extremely important development in 1999 was the Federal Communication Commissions' (FCC's) allocation of significant wireless spectrum, for the exclusive purpose of "protecting the safety of the traveling public" (28). This was the very licensed bandwidth – put to work as Dedicated Short Range Communication (DSRC) – that was later used for the early deployment stages of V2V and V2I. Here was a case of a targeted safety application with immense consequences. The very open nature of ITS allowed this to occur. Did this mean that ITS had now defined itself? Not quite, although it is true that ITS has increasingly focused on the paradigm of connected and AVs (CAVs) – with their national significance – and less on branded initiatives.

3.9.6 EVOLVING SAFETY IMPACT OF ITS

The original impetus for ITS was greater throughput of traffic with less new highway construction and all the while maintaining a safe system. According to FHWA data (29), total vehicle miles of travel in the United States followed the anticipated inexorable upward trend from the start of the ITS era in 1991 for a period of more than 20 years – until the economic recession of 2008. Over the same period, NHTSA's fatality rate per 100 million miles of travel reduced by 34%, from 1.91 to 1.26. This was a positive total outcome in terms of accommodating more and more use of the network and improving the safety of all that travel at the same time.

ITS was never all about safety: it was about greater productivity of movement without building more and more interstates and without compromising safety. But ITS has always strived for safety improvement and has succeeded in that goal, while simultaneously addressing its goals of traffic efficiency and environmental improvement. The starting point for ITS was intelligent highways but rapid advancements in motor vehicle intelligence became more and more important in the long run. ITS has been able to play a useful role in transferring sensor technologies from vehicles to highways, particularly intersections. Rather than a basket of ever-changing technologies, ITS is settling on the connected vehicle and infrastructure technologies of V2V and V2I and is the logical clearing house for widening deployment of CAVs. The predominant concern of ITS would then be safety – crash avoidance to be specific – and ITS has the potential to personify the "safe system" called for in modern road safety strategies.

3.10 THE PHILOSOPHICAL PRINCIPLES OF VISION ZERO

3.10.1 ROAD SAFETY FATIGUE

By the end of the 20th century, road safety fatigue was setting in for government road safety agencies in many countries. After decades of supporting road safety as a vocation, with research, government programs, policy development and regulation, there had been no major breakthroughs. Progress in reducing crash rates was slowing and the total quantum of death and injury remained about the same. It appeared that the world was happy to continue with ever more drivers, more cars and more miles traveled. The popularity of personal road transport was showing no signs of abating and the toll of road vehicle crashes was not proving to be a show-stopper.

Some smaller countries – more nimble of policy – decided on a new road safety paradigm that deviated from long-standing safety gospel. Vision Zero was based on the idea that nobody should lose their life through making a driving error, and that life and limb could not be traded for material or transient things. Without saying so explicitly, Vision Zero decided to end approbation for nuanced measures that permitted more driving without increasing harm (but also maybe not reducing harm). Perhaps this was no longer going to be good enough.

The vocation of road safety had developed a habit of setting targets that involved certain percentage reductions in death and injury metrics, within a certain number of years. Many countries and domestic jurisdictions developed five-year or ten-year strategic plans that parsed how such improvements would be managed. Governments were not willing to commit themselves to major breakthroughs – rather the pace of incrementalism would be increased and closely managed.

With a movement called Vision Zero, a major strategic disruption was contemplated. Proponents saw that too much attention was lately being devoted to "driver error" and were well aware that the whole field of vehicle safety had been extensively addressed throughout the latter part of the 20th century. Safety experts recognized that the potential safety contribution of the underlying road network had been overlooked; they also began to venture into the realm of systems behavior. In the new paradigm, road safety was a shared responsibility of the transportation system designers and the road users. The holy grail was the safe system, and proponents were determined that it was wrong for anyone to be killed on the public road system, thereby paying the ultimate price for making a mistake while driving.

3.10.2 Launch of Vision Zero

In October 1997, the Swedish parliament approved a new national road safety policy, and it was called Vision Zero (30). It stated that the nation's streets and roads would be managed to prevent fatalities and serious injuries. Appropriate strategies would be introduced so that the following principles would be followed:

1. The designers of the system are always ultimately responsible for the design, operation and use of the road transport system and thereby responsible for the level of safety within the entire system.
2. Road users are responsible for following the rules for using the road transport system set by the system designers.
3. If road users fail to obey these rules due to lack of knowledge, acceptance or ability, or if injuries occur, the system designers are required to take necessary further steps to counteract people being killed or seriously injured.

The meaning of "Vision Zero" was that "no foreseeable accident should be more severe than the tolerance of the human in order not to receive an injury that causes long term health loss".

As such, Vision Zero was a set of philosophical principles, rather than a strategy. It was now being made clear that the "designers of the system" were responsible for safety, that they would set rules for using the system and that road users were responsible for following the rules. The idea that system-based rules would drive safety – as expressed in the first and second principles – was a large philosophical swing away from the utilitarian mindset of road safety. But the inclusion of the third principle retained an "if problems persist" element of old safety and its reliance on countermeasures.

Recall that one of the hallmarks of the Nader era was the idea of two collisions. Having done much to ameliorate the second collision – of human and vehicle interior – the safety community was now calling out the first collision. A combination of system design and rules of use would address collision avoidance or at least civilize the collision. Note that the physical design of the system takes many years to address, although ITS could begin to accelerate system design for safety. As for rules of use helping to avoid crashes, conventional traffic rules such as speed limits are too generic to transform crash avoidance. So Vision Zero principles were transformational but may encounter great challenges in application. Certainly ITS, whose introduction coincided with that of Vision Zero, was "on message". But its origins in traffic efficiency would need broadening to seriously address safety.

3.10.3 DIFFICULTIES WITH VISION ZERO

Many had a problem with the "Zero" aspect of the policy and still do. How could professionals seriously undertake such a "mission impossible"? But if we look a little deeper, we can see that the policy was not aiming for zero crashes but for zero deaths. Could the first collision be "calmed" sufficiently to avoid death or serious injury in the second collision? For one thing, crash severity is extremely sensitive to crash speed. So the Swedes tried to make sure that crash speeds in various roadways settings would always be lower. Since the launch of this policy, the number of road deaths per unit population has more than halved. But would other countries with busy, individualistic populaces be able to emulate this example?

Many countries, states, counties and cities have since imitated Vision Zero, in their own environments and in their own way. An impressive number. Aside from the kinetic logic underlying Vision Zero, it also became apparent that there was a new ethical dimension to the policy. Matts-Ake Belin, who was one the architects of Swedish Vision Zero, put it this way:

The problem is the whole transport sector is quite influenced by the whole utilitarianist mindset. Now we're bringing in the idea that it's not acceptable to be killed or seriously injured when you're transporting. It's more a civil rights thing that you bring into the policy.

Yes, but they were also trying to tell the road designers and traffic engineers that they couldn't just stand by and pin the blame for crashes on human error. Just how realistic is this as a policy position?

Not too far away in The Netherlands, the problem of designing safe road systems has been actively pursued throughout the Vision Zero era, without too much mention of the movement itself. They chose to call it Sustainable Safety. In a 2017 publication (31) entitled Designing Safe Road Systems: A Human Factors Perspective, Jan Theeuwes and his colleagues set out the Dutch approach to making the roads sector an active contributor to the elimination of road deaths and injuries, rather than a bystander. They address "self-explaining roads" and traffic systems and set out principles for roads that will instinctively create safer driving. While deeply rooted in engineering, both physical and human, theirs is almost a mystical approach to one of the key precepts of Vision Zero: driver error should not result in loss of life.

These Dutch experts believe that good road design will prevent crashes if driver perception and decision-making are normalized by improving up-front perceptions, avoiding surprises and providing recovery when an error is made. Road managers had developed the habit of instantly changing speed limits in response to traffic conditions. Rather than thrusting signage at drivers, especially of the dynamic kind, the road itself should provide a consistent feel of the appropriate speed of travel. Inadvertent aberrations should not result in crashes but in "graceful degradation". For example, wrong-way freeway entries should be immediately countered with obvious roadway markings on the ramp. A further salutary lesson for road designers is contained in the knowledge that drivers operate with many competing goals, not all of which relate to driving. In other words, road designers should not assume that they are communicating with rational, fully engaged actors.

The question of the interaction between road designers, or road managers, and road users is often glossed over. Under business as usual out on the road, the communication is one-way. The driver receives information, often instructional or mandatory, via signage and more generally from the visual cues provided by the physical elements of infrastructure and roadside furniture. Most of this information needs to be "one-size fits all". It needs to be minimal and crystal-clear. Otherwise road people could be accused of confusing drivers and maybe making things less safe.

One of the road system's biggest engineering contributions to road safety is traffic control at intersections. A high proportion of urban traffic deaths and injuries occur at intersections of major roadways. Such intersections may be controlled by traffic signals or may rely on signage and limited signals representing a variety of stop sign and warning protocols. Signalized intersections have an incomplete safety function because they still allow opposing vehicles to have severe crashes when one attempts to turn across the path of the other. In these cases, both drivers could still be driving legally, within the road rules. A different, cross-traffic, scenario occurs at unsignalized intersections, where one vehicle slams into the side of the other vehicle. This involves major error, or negligence, on the part of at least one of the drivers. Often, the precipitating vehicle – "Vehicle A" – has followed road rules by stopping, but then proceeds legally but foolishly into the path of the legally compliant "Vehicle B".

In both cases, the road designer is trying to talk to the driver on a bad phone line with a serious speech impediment. This type of communication could also be likened to a wholesaler trying to sell directly to a consumer. Think back to William Haddon's insistence that the mid-century epidemic of vehicle deaths and injuries needed to focus on protecting vehicle occupants once the crash has occurred. The automakers were being forced to protect their own customers, with whom they already maintained an unprecedented level of communication, for marketing and brand loyalty purposes. None of this exists in the case of road designers.

Road designers therefore fall back on generic rules for reducing the severity of crashes. This requires curtailing vehicle speeds in certain types of road environments. The Swedish road safety professional Claes Tingvall, often credited as the father of Vision Zero, collaborated with Narelle Haworth of Monash University in 1999 to guide the implementation of Vision Zero in Australia (32). They produced maximum permissible vehicle speeds for the blanket avoidance of fatalities in several scenarios. These scenarios were described in "infrastructure-related" terms of likely collision directions: opposing (70 km/h), side (50 km/h) and same-direction (100 km/h or more). However, this could hardly be called a meeting of the minds of road designers and road users. Recall that we are trying to avoid deaths and injuries, not crashes. The direct proximity of the injuring product to the injuries was the key to rapid progress in improving vehicle design. In the 21st century, this is not the case with road designers, who remain as isolated from the actual deaths and injuries as they ever were.

3.10.4 MORE INTELLIGENT SENSING AT INTERSECTIONS

One group of road designers who have worked hard to reduce crashes at signalized intersections is the traffic engineers. Coinciding with the era of Vision Zero, intersection design has had to cope with more complexity in terms of higher traffic volumes, more turning movements and more micromobility, including pedestrians, cyclists and electrified devices. From the car perspective, these efforts have still not been directed squarely at deaths and injuries but more at crash avoidance. The situation is different with vulnerable road users, where impacting cars can produce crippling injuries at virtually any speed other than zero.

Advances in the detection of vehicles in and around intersections has enabled the signal phases of green, amber and red to adapt to traffic conditions and for other intersections to be connected to improve traffic flow. However, the relentless rise of traffic congestion has meant that smarter traffic signals have to be doubly smart – for traffic flow and for safety. Again, this does not bring road designers much closer to solutions specific to death and injury. The field of ITS – which has also come to the fore in the Vision Zero era – is trying to do something more by having a real-time connection between vehicles and intersections. This has enabled so-called green wave approaches; not only are multiple intersections coordinated but also drivers are given an advisory speed that will minimize their need to stop at successive red lights.

ITS would also like to go further and provide a host of safety advisories, directly to drivers, that use connectivity with other vehicles in the vicinity of the intersection and with the intersection itself. Such warnings of crash risks could address the most violent types of crashes at intersections and elsewhere. For the first time, road designers and their surrogates could speak directly, loudly and clearly to their road users. Despite the efforts of federal and state road agencies and DOTs, this technology still suffers from the "chicken and egg" problem of committed, sunk investment being required by all sectors: roads, vehicles and drivers.

3.10.5 NATIONAL SAFETY MANAGEMENT

Versions of Vision Zero have now become the dominant road safety strategies of the 21st century. A good example is the National Road Safety Strategy (2021–2030) of Australia (33) which posits stepwise improvements in the safe system, with a goal of reaching Vision Zero by 2050. However, Australia's 2021 review of progress found that targets in reducing fatalities and injuries had not been met, with injuries increasing rather than reducing. The new Australian strategy represents a "full-court press" with minimal recourse to philosophy or doctrine. More integrated management by objectives and a renewed focus on data are emphasized. But comprehensive speed management is the main 21st-century approach in evidence, in this and other national strategies, at least for the coming phase out to 2030.

3.10.6 NORWAY'S MITIGATED LIBERALISM

Wherefore the new road safety philosophy based explicitly on ethical principles? Another small, intense country like Sweden – Norway – was an early adopter of Vision Zero as national policy. This decision was made by the Norwegian Parliament at the beginning of the 21st century – in 2000. By 2005, Norwegian safety professionals were coming to grips with their new policy and exactly how it differed from the previous national policy. This required a deep dive on ethics and morality in the transport system because, tempting as it was to continue studying crash and injury statistics, this did not speak to changes in personal freedom, nor the moral responsibility of institutions. Pandora's Box was well and truly open. In a wonderful report by the Institute of Transport Economics (34), Beate Elvabakk provided an exposition of a range of enduring ethical principles and then applied them to the previous safety status quo in Norway – which she termed "mitigated liberalism" – before turning to a comparative investigation of Vision Zero.

Norway's mitigated liberalism was much like other countries' policies that evolved during the 20th century but without discussing the underlying ethics. Elvabakk articulated Norwegian road safety policy in a way that looked much more like a national philosophy. She showed that utilitarianism ruled through the right to move as freely as possible while minimizing the harm in crashes, deaths and injuries. Yes, we have a system made up of components (roads, vehicles and drivers), but the system – the ghost in the machine – could not be professionally managed. That is, the variegated road safety community was not in a position to claim the moral position of "no fatalities". She believed that we had hitherto contented ourselves managing the individual components.

She saw that an exclusive focus on eliminating road deaths equates to virtue ethics and would quickly lead to extreme social changes. But as soon as you include the utility of road transportation, you're back with utilitarianism. So Elvabakk found that Vision Zero was not a totally new ethical construct, but was more of a mix of ethical tools, and was still strongly utilitarian. The morality of road designers, drivers (and other road users) and vehicle manufacturers was not front and center. And our inability to materialize the interactive system was passed over to those who had previously escaped attention – the roads sector.

Elvabakk's 2005 review of Norway's Vision Zero also found that progress had not met expectations. It was also clear that Zero did not herald a new, more ethical, approach because it was still utilitarian in nature. And even within that branch of ethics, it was inconsistent to require the "no trade-off" principle of human life. Perhaps the biggest societal issue with Zero is the moral responsibility ascribed to the transportation system when it remains humanly open and intractable. In Elvabakk's view, we had become accustomed to the standardized elements of the traffic system: "the standardised driver, the standardised environment and the standardised vehicle". And the interplay between these elements was effectively the ad hoc responsibility of drivers.

Vision Zero introduces a distributed responsibility and probably the need for a more closed system and even a "system administrator". This would be counter to the libertarian instincts – part of a different branch of ethics – of Western societies. Elvabakk also points out that virtue ethics are on the rise, asking the question "what is the good life?" Where does road safety sit relative to other life pursuits? Would we want this question to be answered by a government? She warns that we should be careful of dictatorial safety policies given that the demand for driving is a derived demand, not a primary good.

3.10.7 MORALITY OF ROAD SAFETY

Modern national road safety strategies are owned by governments and therefore tend to seek steady improvement rather than transformational change. Government agencies are required to articulate goals, formulate programs and manage progress toward stated goals. A wide range of "management by objectives" methods are being employed, including quantitative metrics and a range of leading and lagging indicators. Such policy professionalism is helping to reduce the harm on our roads but is not capturing the imagination of road users. Safety professionals are dedicated and long-serving, and may be passionate. They are people of good will but not necessarily inspiring and ultimately virtuous. The vast majority of their driving subjects remain unmoved. Something is missing.

Virtue is an extremely high bar and is a life-long habit. Other time-tested strands of ethics could and should also be applied to road safety. For example, deontology, or the following of rules, should be made more explicit and compelling. One of the great contributions of Vision Zero is to remind us that the road transportation system needs to be furnished with ethical principles for avoiding fatalities, not solely for developing countermeasures after the event. Driving within the speed limit will likely not prevent the first collision, although it may reduce the severity of the second collision.

3.11   IN SEARCH OF THE PRACTICAL ETHICS OF ROAD SAFETY

3.11.1   THE SAFE REFUGE OF PROVEN COUNTERMEASURES

Dr. William Haddon Jr. is widely recognized as the father of modern injury epidemiology, as well as the father of scientifically based road safety interventions. In thinking about phenomena that are harmful to humans, he was conscious of the important distinction between the merely known-about and the understood. He gave (35) the example of the South Seas anthropologist Malinowski who wrote that:

It is most significant that in the lagoon fishing, where man can rely completely upon his knowledge and skill, magic does not exist, while in the open-sea fishing, full of danger and uncertainty, there is extensive magical ritual to secure safety and good results.

He was wary of events that were attributed to "extrarational factors" termed luck, chance, accident or fate. This was certainly the case with road crashes, deaths and injuries where the term "accident" was in common usage for most of the 20th century. Regardless of the field of injury or disease under scrutiny, Haddon was on the prowl for countermeasures. In the case of road safety, he set out to overcome ignorance and prejudice. His philosophy was that mechanisms of harm needed to be understood and countermeasures introduced. It was self-evident in his professional field of medicine that finding harm-reducing measures was the right thing to do. He invented the famous Haddon Matrix of harmful agents and hosts to bring rigor to his work and that of his long-standing, powerful – and yes, virtuous – creation: namely the National Highway Transportation Safety Administration (NHTSA).

3.11.2   THE SOCIAL CONTRACT

The ethical principles of road safety have surfaced in many indirect ways over the years. According to William Haddon, "the government has a right and duty to require levels of safety in motor vehicles". Clearly Haddon was motivated by a strong sense of ethics, and his work was part of the age of regulation. Rights and duties of governments would presumably rest on the social contract theory of ethics. John Hobbes was desperate to escape from the state of nature, where our lives would be "nasty, brutish and short". He thought we all needed to sign over some of our freedoms so that we would be protected from the worst impulses of fellow citizens. Enter the social contract, although still not well articulated in the case of road safety.

America's democratic system of government, and a highly pluralistic society, mitigated against giving much power over everyday lives to the government. One of Haddon's antagonists in law-making for road safety was the Republican Representative Bud Shuster, who became the chair of the U.S. House Committee on Transportation and Infrastructure, and was a well-known opponent of the airbag. He vehemently opposed legislation that "chips away needlessly at our individual freedom". His arguments invoked important ethical principles – those of liberty. John Stuart Mill, whose utilitarianism has increasingly struck a chord in the modern world, was adamant that the government should not intervene unless "mayhem is certain". For Mill, even the likelihood of harm was not sufficient to justify government interference – it had to be absolutely certain.

3.11.3   NHTSA's Expanding World of Crash Countermeasures

Over the years, NHTSA and many other entities involved in road safety doubled down on the search for countermeasures by adding layers of influence to the Haddon Matrix: cultural, institutional, interpersonal and intrapersonal. The system of safety countermeasures expanded well beyond Haddon's closed, professional, medical domain to include significant elements that were not professional and were not subject to medical ethics. The vast cohort of vehicle owners and drivers for one. Consideration of the right actions became much more complex and fraught with dilemmas and contradictions.

Always remaining authentic, Haddon refused to assume that drivers could be relied upon to take the right action, such as putting on a seat belt. He was very skeptical about the total avoidance of crashes, as the road transportation system was evolving. His professional perception of the right thing to do – to protect vehicle occupants when crashes do occur – served the cause of road safety for decades. But eventually it looked like new technologies might also help drivers to avoid crashes. And the right path to follow was not professionally obvious. The ethical calculus was changing. It had been one thing to force automakers to engineer occupant protection into their vehicle interiors, but coercive laws directed at driver behavior brought up many new considerations.

3.11.4   Bedrock of Utilitarianism

The ethical principles of utilitarianism found a grateful home in the worldwide practice of road safety. Often ascribed to John Stuart Mill in the mid-19th century, utilitarianism teaches us that the consequences of actions alone determine whether those actions are right. In safety, the consequences are usually thought of as two sides of the same coin: the happiness created by driving and the harm caused by crashes. Note that happiness might be too strong a word for some modern economists, and the term utility would suffice in order to convey the economic imperative of transportation. Ethical practice implies that actions will be assessed for the greatest happiness of the greatest number and the least harm caused. In road safety, the total number of fatalities per unit population – and many variants – is the ultimate harm metric. Many are uncomfortable with quantifications of death, but the metric of fatalities per unit distance traveled is still widely used and gives a sense of both sides of the safety coin.

3.11.5   Driving Freedom

The good that results from road transportation, and driving in particular, is not measured numerically. A few things may be expressed in a dollar value, such as access to employment and retail services. But there are many intangibles that cannot be ignored. The author Matthew Crawford speaks from experience of the zeitgeist of driving. In his book Why We Drive (4), he travels all the way up the scale of freedoms. Driving not only provides unplanned, unscheduled mobility on the spur of the moment but also points to our need to learn and maintain competencies and agency in the face of risk. He believes that the safety movement wants to take all of this away.

If we go back to Tocqueville's first impressions of America in 1832, we clearly see the biggest two-sided coin of them all: social equality and individual freedom. Coming from the totally unequal French society of the early 19th century, he was acutely conscious of America's enviable point of social departure: political democracy. At first glance, transportation and driving have managed to remain remarkably equal for almost all citizens. But a century of government regulation has increasingly eroded an individual's liberty behind the wheel. Re-reading John Stuart Mill's ethical admonitions against government interference in the mid-19th-century era of roads and railways, one is struck by just how many bags no longer contain cats and barns lack horses. Driving is equal but is also loved more for its freedom, and that freedom is constantly being eroded. This is not ethical. Yes, utilitarianism is supposed to deliver a large volume of happiness and usefulness and as little harm as possible. But it is also meant to preserve liberty across the large swath of society that wants to drive.

If we are willing to think of roads and driving as a system with inputs and outputs, then the utilitarians have long been having a field day with measuring the outputs. Let's call these back-end ethical considerations – lots of statistics. But what of front-end ethics? What contracts may have been struck or implied? How much authority has been passed over to the government? What rules are applicable? Actions would need to be assessed against such rules. Higher up the scale of morality, what actions are always considered to be good, or in a somewhat different sense, right? – "the right thing to do". Then, would Aristotle feel that we give individuals enough scope to develop and exercise virtues, and each represent "an end in themselves" – never a means to an end. Who or what could serve as a "stable and general" presence, a beacon on the hill for such an alarmingly practical application of ethics?

3.11.6 THE ARC OF NHTSA's SAFETY LEADERSHIP

Would NHTSA represent such a virtuous organization, especially in its earliest days? This would of course be a tall order. But NHTSA was able to evolve in its approach to harm reduction and move beyond countermeasures to include driver behavior. As the vehicle was made safer and safer to be in, the crash statistics pointed the finger more at driver error. NHTSA's motivation was moving beyond occupant protection to dream the impossible dream of crash avoidance. However, the motivational and rational purity of action evident in occupant protection did not transfer cleanly to crash avoidance. To appreciate the problem, we need only consider that the human was now part victim and part villain. The single-vehicle methodology of crash consequences – wherein so much more is known about the actions of Vehicle A than those of Vehicle B – was no longer sufficient in the majority of serious crashes; and the police reports that form the basis of crash knowledge say little about "Driver A", let alone "Driver B". Vehicle A remained the anchor of crash knowledge, which continued to be obtained after the event.

What groups other than NHTSA had the motivation and ability to take good actions in road safety? Certainly police forces and highway patrol squads had long and admirable first-responder and enforcement experience. And they had made a tremendous contribution by collecting and archiving crash data. Almost from the earliest days of powered vehicles replacing the horse-drawn, there was a primary thirst to find out the causes of crashes and attach blame for bad behavior. Such investigations, and resulting explanations, often only proceeded as far as the driver. It was pretty easy to build on the foibles of the driver, particularly once it became obvious that crash avoidance was extremely challenging as a policy.

Within a long-standing atmosphere of human frailty and culpability, governments were on safe ground in exhorting everyone to "drive safely" – this would certainly be the right thing for governments to do. But how effective would it prove to be? Worrying exposes of drivers being inattentive behind the wheel moved on to revelations about drivers being distracted. Drivers would even be so exhausted that they would literally fall asleep. Crash investigators had trouble figuring out whether to code this as "drowsy driving" or "asleep". It would be the right thing to protect innocent drivers from risks caused by all of these previously unknown forms of unsafe driving, as well as an increasing amount of aggressive and irresponsible driving. NHTSA worked hard to develop a better understanding of unsafe driving and the notion of risky driving emerged. NHTSA's FARS and other large sources like the GES were interrogated to glean information about the many adverse situations that drivers were placing themselves in and the errors, oversights and derelictions that then completed the crash narrative. But risky driving remained a catch cry more than an analytical tool.

NHTSA's pioneering research in the field of human factors had taken a benevolent approach to people who drive. Under NHTSA's gaze, engineers were determined to design vehicles that were conducive to human control and would minimize challenges to safe passage in roadway traffic. Now it was becoming obvious that such good intentions were being abused by at least some drivers. Again NHTSA stepped into the unknown and carried out large naturalistic driving studies that excelled at measuring the mechanics of safe passage. But actual crashes are so rare on a per-mile basis that the adverse rush of events erupting in a collision were only seen in a few cases and these of low severity. The practicalities of mounting such studies meant that they could provide a much better understanding of conflicts, or near-crashes, than actual crashes. But would such understanding carry through to crashes of high severity?

We may well believe in a continuum of risk that can be expressed as traffic conflict, near-crash event, property-damage only crash, minor injury, serious injury and finally fatality. Starting at the latter end of the severity spectrum, if we move to the left through stages of less severity when do we stop talking about countermeasures? At some point, risk reduction no longer follows the utilitarian ethic and becomes a matter of good will. Does it make sense to try to develop countermeasures – virtual countermeasures – to traffic conflicts? Therefore, does our precious belief in the concept of countermeasures have its limits?

3.11.7 THE EUROPEAN PHILOSOPHY OF VISION ZERO

Finally, in the last decade of the 20th century, a new strain of road safety intervention emerged and pushed back against the sterile blaming of drivers. Vision Zero originated in Sweden, a professionalized country otherwise known for making safe and innovative cars, and for entertaining government actions that erred on the side of social leveling. Right out of the gate, Vision Zero was heralded as a more ethical take on road safety because – for the first time – it placed onus on road designers and managers. It was also extremely philosophical in that crashes were no longer branded as driver error but as "system failures". This led to the idea that the system should be a lot more forgiving and be able to convert catastrophic driving failures into manageable driving failures. It featured two moral declarations: that human life would not be "traded off" with the benefits of driving and that nobody should die on the roads when they make a driving error.

It must be said that the goal of zero fatalities changes everything from an ethical perspective. It is a strong front-end statement of virtue, what is right and what is true – thou shalt not kill. If it was left at "we aim to halve fatalities", it would have remained a back-end utilitarian philosophy driven by the numbers. A virtuous goal needs a virtuous organization to go with it. Who in the whole field of transportation would be in a position to make a commitment to zero fatalities? Critics have pointed out that it may be appropriate for a closed, well-resourced, professional society – such as medical practitioners – to adopt such a principled goal. But transportation is far too open, loose and unprofessional to do so. The 2005 Norwegian review of Vision Zero (34) provides a synopsis of these ethical non-sequiturs.

Regardless, Vision Zero was adopted in many countries and jurisdictions. But it was not easy to translate the philosophical into the engineering. Road and transportation agencies continued to apply every available measure and countermeasure to improve road safety. The ultimate test still remained the key crash and fatality statistical measures, and the philosophy therefore remained staunchly utilitarian. Promoters of Vision Zero went further into the realm of moral conviction and stated that fatalities and injuries could not be "traded" for the freedom to drive at will. However, there is no evidence that Americans changed their ways in order to take moral responsibility for the avoidance of harm to others. In fact, many jurisdictions interpreted Vision Zero as a "stretch goal", in the parlance of business books. Brought back to earth, Vision Zero placed a much greater focus on speed, and less speed would always be better. Although some jurisdictions, such as Australian states led by Victoria, directly messaged the driving public about human beings' low tolerance of impact speed ("Speed Kills"), most jurisdictions were hoping that regulators would take action for road rules prescribing lower speeds in general.

There is another problem with the fictional notion of trading the lives of those killed in road crashes with the thrill or utility of freedom to drive. From an ethical perspective, there is a huge difference between acting in the knowledge that lives could be lost and taking an action that will directly lead to someone's death. The trade-off statement sounds more like the latter than the former, and the latter type of trade-off is institutionally impossible. We know that the portfolio of transportation is conducted for the overwhelming benefit of society and in the knowledge that some lives will be lost. A part of that portfolio – a progressive mechanism we call road safety – works to minimize harm to humans. Separate mechanisms exist to address other virtues important to the transportation portfolio, such as mobility, sustainability, traffic efficiency and social cohesion. Road safety cannot control these other mechanisms and is not in a position to trade off one transportation virtue with any other. While the trading of lives would indeed be a monstrous idea, it never happened and never could happen.

To the extent that the portfolio of transportation is being fashioned as a virtuous enterprise, it works simultaneously on a number of distinct societal values. Aristotle told us that a virtuous person does not try to maximize all of their desired virtues but seeks the Golden Mean with each. That is, virtue turns into vice if it is done to excess, and it also turns into vice when it is insufficient. In the case of the Golden Mean of safety, we must avoid being over-protective or stifling, and we must also be sure to avoid being detached or irresponsible. We should not be seeking a straight-ahead maximization of safety. But many interpret Vision Zero that way. And that is the linear assumption behind the trade-off analogy: that safety could be minimized in order to maximize another virtue – perhaps some notion of personal freedom or economic advantage. If we accept that transportation should contain a number of parallel values, then they should be sufficiently independent to allow each to seek its Golden Mean.

The subject of practical ethics for the purpose of road safety – an attribute of the relentless but informal road transportation system – has largely remained unarticulated. Ethical references arise in every nook and cranny of this public–private personal–industrial pageant but only as required. Nobody can really say what a safe system would be, or how it could be curated or its arrival hastened. It would certainly be an ethical hybrid, with good intentions on the part of individuals and institutions and good system outcomes brought about by exemplary professionalism, good system administrators and authentic behavior all round.

Contained within that unprescribed "safe system" would be sub-systems of expertise – some more closed and professional than others. This would include medical expertise, automotive engineering expertise and traffic engineering. In the worlds of academia, technology, economics and culture, transportation itself is not viewed as a field totally worthy of fundamental investigation and of discovery. It is the low-hanging fruit for any of them to test their ideas – because it is seen as data-rich but knowledge-poor. If you were to count departments of transportation among the world's top 100 universities, the answer will be zero. This probably reflects transportation's status as a non-primary demand in our economy. Transportation needs to stay in its lane and so does transportation safety.

3.11.8   SAFETY AS AN ATTRIBUTE OF THE TRANSPORTATION SYSTEM

Though the term "transportation system" rolls off the tongue, it is not really a system, certainly not a closed professional system that could be held to account on matters of principle. Road transportation, covering movement of both people and goods, represents a large chunk of GDP and is eternally indispensable. It is a large and boisterous portfolio, with loose ties – difficult to visualize and comprehend. The private side of people movements includes vehicles, fuel, insurance, finance, for-hire services, tolls, fees and parking. On the public side, various cost centers – many of which are outsourced – include road design, construction and maintenance, traffic control and enforcement. Freight movement covers both own-account and for-hire carriers, trucks and trailers, a huge component of fuel and drivers' wages, depots, terminals, yards, warehousing and rest areas. Freight movement is subject to many kinds of regulations and fees and charges levied by states. Because commercial truckers use the public road system, there is a general political commitment to recovering costs of design and construction to accommodate their unusual weight and size, as well as wear and tear.

It is difficult to contemplate the ethics of such a portfolio without understanding the ties that may act to bind all of these industries together. The influence of governments – federal, state, city and local – is obviously felt by all players but in different ways. The U.S. DOT, which did not exist until the mid-20th century, incorporates separate "modal" administrations. The most prominent of these are NHTSA and the FHWA, along with the Federal Motor Carrier Safety Administration (FMCSA) and the Federal Transit Administration (FTA). Other federal departments also have an important influence, including the Department of Commerce where the EPA is located. The supreme oversight agency of the U.S. Congress is the Government Accountability Office (GAO). At the state level, all have DOTs that cover roads and bridges, vehicle registration, traffic, safety, and fees and charges. In many states, only a minority of roads are "owned" by the state, and the majority are managed by counties and cities. Driver licensing is carried out by state police and enforcement by locally organized highway patrol departments.

Private companies range all across the road transport industry and make common cause as required, often for lobbying or policy influence related to pending regulation or changes in economic conditions. The vehicle manufacturers and their various industry associations – both light and heavy vehicles – are clearly important players and constitute regulatory cartels. Their continuous stream of transactions with NHTSA, FMCSA and others represent a closed, professional system with clear principles of conduct and process. This really constitutes two systems: one to handle vehicle safety and one to handle vehicle emissions.

The other major federal player, FHWA, has more of a standards-setting brief than a regulatory role. FHWA deals with the DOTs of all 50 states in setting standards for roads and traffic. The states caucus together in the form of the AASHTO. This set of activities involving FHWA, AASHTO and DOTs is extensive, continuous and professional, and closed. As a system, it lacks the cause-and-effect urgency of the two vehicle systems.

Organizations like the American Automobile Association (AAA), itself a network of federal and state organizations, represent vehicle owners and drivers. Along with many other industry and professional organizations, the AAA has sufficient status to be consulted on government investigations and rulemaking. These activities are too diffuse to rise to the level of a system.

There are many other ways in which parts of the transportation portfolio interact, but one of the most valuable is transportation research. The National Academies of Science, Engineering and Medicine include the Transportation Research Board (TRB) which was presciently established in 1920. TRB is mainly focused on the roads and traffic part of the portfolio, rather than vehicles. Among many benefits, it not only provides the state DOTs with an R&D capability but also identifies major national challenges in transportation and hosts an informed, professional exchange that rises to an academic level. This focal point for research has most of the characteristics of a professional system but is not closed. However, there are clear qualifications for taking part.

If we think through the concept of transportation as a portfolio, it is a portfolio of action and money based on a broad set of industries. The portfolio exists to serve the consumers, or users, who pay for everything. The portfolio operates via an interconnected constellation of franchises. These include vehicles, roads, energy, traffic, regulators, insurance, mobility, freight and enforcement. This boat is loaded with society's weight of expectations. Attempting to keep it afloat are a range of progressive mechanisms that each rely on freedom of association and information exchange, including: safety, energy efficiency, traffic efficiency, social coherence and economy. Each of these mechanisms contains a number of professional sub-systems which may have system administrators and virtuous overtones. Our boat also receives some help from the air currents of government policy, social contract (fairness, ethics), political discourse, professional bodies, social media, academia and science and technology.

Does it make sense to consider the philosophies of road safety in isolation from transportation as a whole? Clearly not – safety exists alongside a number of other important values in the portfolio of transportation. Even if transportation lacks a clean taxonomy and a set of universal principles, it is the biggest personal action set embraced by the people of America. And its biggest money pit. Endless circulation of people, machines, commerce and culture; in service of people's most important needs and desires. And it continues, growing and evolving year after year, decade after decade, century after century. It is far too active, hungry and pragmatic to stop and understand itself or fix its problems. Nevertheless, safety has the distinction of being transportation's conscience and memory; the only one in the family that's been to college; the only one who can figure things out from first principles. Road safety has a rich philosophy but keeps it under wraps so it won't embarrass its siblings.

REFERENCES

1. Haddon, William, Suchman, E.A. & Klein, D. (1964) *Accident Research: Methods and Approaches*. Harper & Row.
2. Moynihan, Daniel P. (1959) Epidemic on the Highways. *The Reporter*, April 30, 1959.
3. Gladwell, Malcolm. (2001) Wrong Turn. *The New Yorker*, June 11 2001.
4. Crawford, Matthew B. (2020) *Why We Drive: Toward a Philosophy of the Open Road*. HarperCollins Publishers.
5. Wurts, Richard & Appelbaum, Stanley (2013) *The New York World's Fair 1939/ 1940: In 155 Photographs*. Dover Publications.

6. Fairbank, Herbert S. (1939) *Toll Roads and Free Roads*. Bureau of Public Roads.
7. Malcher, Fritz (1935) *The Steadyflow Traffic System*. Harvard University Press.
8. McClintock, Miller (1927) The Traffic Survey. *The Annals of the American Academy of Political and Social Science*, Vol. 133, No. 1.
9. Mill, John Stuart (1859) *On liberty*. Cited in Macleod, Christopher, "John Stuart Mill", The Stanford Encyclopedia of Philosophy (Summer 2020 Edition), Edward N. Zalta (ed.). https://plato.stanford.edu/archives/sum2020/entries/mill/.
10. The Policy Circle (2021) *Government Regulation*.
11. Washington Post (1997) *Mercedes Benz Recalls A-Class After it Flunks Moose Test*. Nov. 12, 1997.
12. National Highway Traffic Safety Administration (2018) *NHTSA's Safety Performance Measures Selection Criteria. DOT HS 812 628*.
13. Evans, Leonard (1999) *Traffic Safety and the Driver*. Science Serving Society.
14. Rumar, Kare (1999) *Transport Safety Visions, Targets and Strategies: Beyond 2000*. European Transport Safety Council.
15. Hakkert, A.S. & Braimaister, L. (2002) *The Uses of Exposure and Risk in Road Safety Studies*. Dutch Institute for Road Safety Research (SWOV). R-2002-12.
16. Borkenstein, R.F. (1964) *The Role of the Drinking Driver in Traffic Accidents*. Department of Police Administration, Indiana University.
17. Hansson, Sven Ove, "Risk". The Stanford Encyclopedia of Philosophy (Summer 2023 Edition), Edward N. Zalta & Uri Nodelman (eds.). https://plato.stanford.edu/archives/sum2023/entries/risk/.
18. Organisation for Economic Cooperation and Development (OECD) (2013) *Road Safety Annual Report 2013*. OECD.
19. International Transport Forum (ITF) (2024) *Transport Safety. Road Injury Crashes, Fatalities and Injuries*. https://doi.org/10.1787/g2g55585-en (accessed on 8 October 2024).
20. National Highway Traffic Safety Administration (2023) *The Economic and Societal Impact of Motor Vehicle Crashes, 2019. DOT HS 813 403*.
21. McCarthy, Niall (2019) The U.S. States Pouring Money Into Their Highways. *Forbes*, Aug. 23, 2019.
22. Transport Accident Commission (TAC) (2023) *TAC Annual Report 2022–23*. www.tac.vic.gov.au/__data/assets/pdf_file/0007/787399/2022-23-TAC_Annual_Report-Web.pdf
23. IBISWorld (2024) *Automobile Insurance in the U.S. – Market Size (2004 – 2030)*. www.ibisworld.com/industry-statistics/market-size/automobile-insurance-united-states/
24. Logue, Kyle D. (2019) *Should Automakers be Responsible for Accidents?* Cato Institute, Regulation, Spring 2019.
25. Forbes Advisor (2024) *Car Insurance Facts and Statistics 2024*. June 11, 2024.
26. IVHS America (1992) *Strategic Plan for Intelligent Vehicle Highway Systems in the United States*. IVHS America Report No. IVHS-AMER-92-3.
27. Congress, Nita (1994) The Automated Highway System – An Idea Whose Time Has Come. *Public Roads – Summer 1994*, Vol. 58, No. 1.
28. Federal Communications Commission (FCC) (1999) *FCC Allocates Spectrum in the 5.9 GHz Range for Intelligent Transportation Systems Uses*. Report No. ET 99-5. Oct. 21, 1999.
29. Federal Highway Administration (FHWA) *Monthly Traffic Volume Reports*. www.fhwa.dot.gov/policyinformation/travel_monitoring/tvt.cfm

30. Kristianssen, Ann-Katrin, Andersson, Ragnar, Belin, Matts-Ake & Nilson, Per (2018) *Swedish Vision Zero Policies for Safety – A Comparative Policy Content Analysis. Elsevier, Safety Science*, Vol. 103, Mar. 2018.

31. Theeuwes, Jan (Ed.) (2017) *Designing Safe Road Systems: A Human Factors Perspective.* Routledge, CRC Press.

32. Tingvall, K. & Haworth, N. (1999) *Vision Zero – An Ethical Approach to Safety and Mobility.* Monash University Accident Research Centre.

33. Commonwealth of Australia (2021) *National Road Safety Strategy 2021–2030.* www.roadsafety.gov.au/sites/default/files/documents/National-Road-Safety-Strat egy-2021-30.pdf

34. Elvebakk, Beate (2005) *Ethics and Road Safety Policy.* Norwegian Institute of Transport Economics. Report 786/2005.

35. Haddon, William (1973) *Energy Damage and the Ten Countermeasure Strategies.* Insurance Institute for Highway Safety.

4 The Evolution of Mobility in the New Century

How much scope for improvement does road safety have? Nobody is standing still. New technologies beckon but countermeasures are slowing down and new forms of harm are increasing. More concerted application of knowledge derived from past crashes is not enough. Road safety certainly built an impressive home for itself in the latter part of the 20th century. But it is located in a rough neighborhood. Transportation itself is a complex system; the enterprise of road safety has demonstrated an ability to remove many undue risks within that system. In certain cases it has succeeded in changing the system itself. A prime example is motor vehicle safety regulation. This clean exercise in virtuous intervention paved the way for regulation of vehicle energy and limiting the release of greenhouse gases (GHGs) into the atmosphere. The notion of conforming vehicles spreading improvement right across the country – for the life of the vehicle – is compelling.

The first two decades of the 21st century have been bound up with positioning new paradigms of safety including connectivity and automation. These are not countermeasures, nor even technologies. They are technological for sure but are based on systems of beliefs. Once again safety has the opportunity to lead and improve the transportation system. System technologies that prevent collisions lie before us. They are in accord with the practical ethics and bold principles of Vision Zero but require change in the driving system. They also require more widely shared beliefs across the enterprise of road safety.

4.1 THE COMPLEX SYSTEM OF GROUND TRANSPORTATION

4.1.1 An Evolving National Portfolio of Societal and Industrial Import

Transportation is often referred to, somewhat loosely, as a system. But it's more like a portfolio, in the sense of business and finance, but also in the sense of a catalog of services, both retail and wholesale. And it is conducted on the nationwide public infrastructure commons – highways, roads, bridges, streets, sidewalks and plazas. So who is in charge when it comes to the utility and civility of the infrastructure commons? Some commentators believe that important societal goals are being neglected in order to keep the wheels turning and support the economy. If transportation is indeed a

DOI: 10.1201/9781003483861-4

system it should be possible to make wholesale changes or adjustments to better meet societal goals.

The national enterprise of transportation is unique and complex because it reflects the American way of life. Transportation is one of the most important national domains in support of the good life, and this sets it apart from other national pre-occupations like defense, security and health. Transportation is not defensive; it is the pro-active enabler of many aspects of society, the economy and national resilience. Economists and planners say that transportation is a derived demand, signifying that it does not stand in its own right but is generated by the demand for something more fundamental, like food or energy or entertainment.

In 2021 transportation services accounted for a full 8.4% of GDP (1). And the Dow Jones Transportation Average is the oldest index on the U.S. stock market, predating the Dow itself. In one of the most systematic national classifications, as applied by the Cybersecurity and Infrastructure Security Agency (2), Highway and Motor Carrier is only one of seven key national transportation subsectors, containing 4 million miles of roadways, 600,000 bridges and 350 tunnels. But we need to add more sub-sectors – the 278 million registered vehicles and the 233 million licensed drivers – to come up with the basic pieces of the road transportation system. Pulling all of these elements together, the U.S. transportation and warehousing sector comprises more than 4 million private businesses and had approximately 11 million employees in 2021.

The term transportation is often used in a vocational sense and in alignment with the U.S. stock market view. This would place large trucking companies and parcel services at the top of the tree, each with typical market capitalization of $100–150 billion. However, this view would not include automakers for one, who are classified under the industrial market sector. Companies like GM and Ford have market capitalization of approximately $50 billion, and Tesla currently beats them all at $435 billion. The transportation portfolio includes all of these sectors, plus energy companies, road construction companies and insurance companies. This large and powerful fraternity operates transportation services and generates the transportation economy. The portfolio of transportation exists without a board of directors, a CEO or an advisory board and touches the daily lives of virtually every citizen. But it is also a support act, not the star of the show.

If we did use the term transportation system it would connote nodes and links and a sense of supply and demand. This is mostly too much of a stretch; the part of the portfolio that most resembles a system is the freight sector. This sector provides enduring shape to the larger portfolio because all of the 50 state departments of transport engage in regulatory dealings with that sector. Over a long period, the complex system of transportation is evolving and is becoming more system-like. This is reflected in professional language. The term transportation tends to be used more in a vocational sense, such as the movement of goods, and the term mobility is used to refer to personal mobility. This distinction makes sense because goods transport does represent a system of supply and demand, while personal transportation is more discretionary, is carried out in support of a range of other primary purposes and might even be fun.

4.1.2 THE DAILY MODES OF TRANSPORTATION

Putting the essential sectors of roads, vehicles and drivers to work, transportation of people and goods might be described in terms of trips that have origins and destinations. On a daily basis, these trips might be chained, with several links in a person's work–life cycle. If some of these links use public transit, trips become multi-modal. Americans make more than 1 billion personal trips every day, with average length of 40 miles. Almost half of daily trips are for shopping and errands and only 15% for commuting. The aggregate of all this driving is reported as annual Vehicle Miles of Travel (VMT). In a parallel universe that uses the same infrastructure commons, America moves over 50 million tons of freight every day, and the average trip length is about 200 miles. This commercial set of tasks is carried out by professional drivers, specifically licensed, trained and monitored so that they don't drive too many hours day after day.

The general driving public are not composed of professional drivers. The idea of such drivers diligently carrying out A-to-B driving tasks organized rationally and carried out expeditiously and economically – and safely – does not apply in the majority of cases. Car owners and drivers have limited knowledge, and most do not make good use of the assistive technology that is now widely available in vehicles. Most trips are not well-planned. Drivers have minimal training, and once they pass their license test are out on the road learning from experience. Traffic demands on drivers vary greatly and some drivers become extremely skilled, but there is no standardization of driving. They are all part of a lowest-common-denominator set of actions, influenced by road rules and limited enforcement by the highway patrol. Personal mobility is not a professionalized system; it is a conglomerate that seeks its own level. It cannot be managed or planned.

So far, we see that the American portfolio of transportation assets and participants carries out two aggregated tasks, one of social significance and one of economic significance. While both are subject to many rules and restrictions at the granular level, they have minimal top-down control. They are probably better described as ecosystems, or associations, that interact through a common language and set of values. They are resilient to disturbances and are contrived to keep functioning but do not resemble systems designed for certain levels of performance, or capable of optimization.

All transportation activities are bound together with infrastructure in the form of roads and vehicles. Roads are standardized to a certain level but are also franchised in that they are built, maintained and operated by numerous state and local government agencies and their expert contractors. Vehicles are highly standardized, represent large personal investments and have a unifying influence on the way people drive. Traffic control is a critically important professional franchise and is widely distributed throughout the states, counties and cities. Other operational arms within the transportation portfolio include energy, regulation, enforcement, mobility on demand and car rental. Transportation is therefore executed by franchises that operate in a distributed manner, offering services that are very similar but not quite the same. This notion of franchising is important in trying to understand the complex system of transportation.

All of the tiles on the roof look similar, but the house could not be regulated against leakage.

It should by now be apparent that there is no comfortably settled language for professional discussion of ground transportation. Its primary nouns we find ourselves using shift with the context: system, portfolio, enterprise, industry, mode etc. There is a definite pull to use the term "transportation system", but we are aware that this could be wishful thinking. A more apt representation would be "transportation complex-system"; however, we will continue to use the term transportation system but loosely.

4.1.3 ORDER IN THE TRANSPORTATION SYSTEM

When something goes radically wrong in the course of their driving, most drivers fall back on the insurance industry. This large industry pulls in about 80% of drivers because they don't like to risk their expensive personal vehicle – which may even be named, like a member of the family – out in a sea of strangers behind the wheels of their own vehicles and who could do them a lot of damage. Despite the ubiquitous hold of insurance across drivers and vehicles, the highly competitive nature of the industry means that it could not be viewed as a unifying or systematizing factor.

How well is the agglomeration of transportation able to support a multiplicity of American citizens' essential life-purposes? The nation expects even a servant of many masters to display qualities that include safety, efficiency, economy, social coherence and sustainability. These virtues are the ties that bind a portfolio of convenience and create our transportation system. They necessitate networks of association and influence, as well as professional cadres. It is almost 200 years since the young Frenchman Alexis de Tocqueville highlighted just how effectively common goals could be achieved by freely associating Americans.

The efficiency of use of the infrastructure commons is probably the transportation systems' first duty of care – even before safety and sustainability. It requires a balance between relentless and stressful traffic conditions, with the flow on the edge of breaking down, and a perfectly good roadway being idle, a white elephant. As traffic density increases, efficiency increases. Quite suddenly a point may be reached where some drivers are uncomfortable with short headways and slow down. At the other extreme, nobody wants to see infrastructure with too few vehicles on it – it tends to look grandiose and maybe a waste of public money. Efficient use of the infrastructure is part of transportation's grand supporting role for society. This is orchestrated by the state road and traffic agencies, with support from local bureaus and the U.S. DOT's Federal Highway Administration (FHWA).

There is a difference between freight movement and people movement when it comes to the care and feeding of the transportation industry. State road and traffic agencies have a huge say in the "what, where and how" of heavy truck operations on the commons. As does the U.S. DOT via its Federal Motor Carrier Safety Administration (FMCSA), with duty of care for the safety of motor carriers and professional truck drivers.

When it comes to personal vehicles, the tone is mainly set by the U.S. DOT's NHTSA setting safety standards. Another lesser known party – not often considered in safety circles – is the Environmental Protection Agency (EPA) who set the agenda for vehicle classes and footprints. These class definitions are essential to setting and enforcing quantitative rules for fuel consumption and emissions: bigger vehicles are allowed to use more energy, and emit more, than their smaller cousins. The vehicle footprint is a simple measure of the area between the four tires of the vehicle. Manufacturers take care to design the chassis of a particular model – be it sport utility vehicle (SUV), truck or sedan – so as not to incur fuel economy or greenhouse gas (GHG) penalties. According to the EPA (3), significant changes are occurring across the board. Since 2004, there have been notable increases in fuel economy (32%), horsepower (20%), weight (4%) and footprint (5%). Note that these are national fleet averages; they include consumers exercising preferences for larger vehicle classes, as well as increases within vehicle classes.

Transportation is defined in part by its key gatekeepers and unifying influences. These are the U.S. DOT, through NHTSA, FHWA and FMCSA, automakers, the EPA and the numerous state road and traffic agencies. Each plays an important part in the traffic that swarms across the commons. This is done in a way that tries to limit inconvenience to the users who rely on it for their daily life-enhancing purposes. If one spends a day in Navy Yard in the various halls of the U.S. DOT agencies, there are few cultural disconnects. However, the environments of the U.S. Department of Energy (DOE) or the EPA represent different cultures.

The use of the infrastructure commons for direct social purposes, rather than mobility, is distinct from the transportation conglomerate but is closely affiliated. It is no longer a mere servant of people's real primary purposes. After all, a "great place to be" is a primary good. Its importance has been under rated but has not been fully researched or quantified and is difficult to articulate. Certainly the idea that roads and streets do not simply belong to cars has taken hold. This has resulted in urban designs that are intended to calm vehicle traffic and invite newer modes of electrified micromobility, as well as walking and biking. Such developments are a work in progress but definitely impact the movement of people and goods. With more and more pedestrian-only zones, bike lanes and mixed-use urban designs, strong values associated the urban commons, and the local commons, are influencing the evolution of our ground transportation system in the 21st century. These new sets of objectives – and the city designs that materialize – are under the control of new mobility players. The city administrations involved are numerous and widely distributed; they are much more directly influenced by their citizenry than are federal agencies.

4.1.4 CONGESTION AND DISCORD IN BIG CITIES

Despite the efforts of city planners, traffic congestion remains the scourge of many big cities, and urban innovations have not yet brought about dramatic improvements. This is surprising in relation to findings of the 2019 New York City Mobility Report (4), showing that 64% of trips in the city were by "sustainable mode". Namely, walking, biking, bus or subway. In some parts of the central city only about one third of vehicles on the street are private cars. There has been a major long-term decline in car trips into the Central Business District (CBD). Between 1995 and 2017, sustainable mode trips into the CBD increased by one third and trips by cars and trucks declined by almost one quarter.

Such changes are taking place in cities all over the country and each in its own way. The long-term effect is to increase livability of cities and to improve sustainability, broadly defined. However, reduction in private car use is accompanied by increases in for-hire car use and home deliveries. The stubborn streets of New York City remain heavily congested. Average vehicle speeds – a broad surrogate for congestion – show a long term reduction and remain well below 10 mph for buses. In other words, congestion is getting worse. Similar scenarios are playing out in other large cities, perhaps to a lesser extent.

City administrations make many decisions affecting mobility, including changes aimed at enhanced urban livability and sustainability. These organizations therefore represent an important franchise in our portfolio of transportation. We must remember that city agencies have many responsibilities beyond transportation and they should not be expected to lead the way in its overall philosophy and practice. But they have been influential in the increasing use of the term "mobility" which brings with it a demand for multiple options when moving around the city. And because city agencies are more interested in places than in connecting places, they bring a different sensibility to the overall transportation portfolio and challenge long-standing assumptions about good transportation practices.

The city administrators who seek to enhance values of place also bring an impetus for innovation. New street layouts that force conventional traffic to share more space with practitioners of micromobility – small personal machines – and active mobility also promote sustainability. The safety impacts of such changes are not always clear, and the traditional utilitarian ways of quantifying injuries and deaths are not applicable in many cases. Big cities are trying hard to become a better place for close-in residents but may not be doing much for those that need to drive in and out of the city or travel across the city. Commuters would be hard-pressed to ignore the seeming inevitability of gridlocked traffic.

4.1.5 EVOLUTION IN THE POSITIVE

The practice of transportation is a very broad church, and it would be unreasonable to expect it to come together as a right-thinking whole. If tragedy is just around the corner, we should be looking very closely at the incessant traffic of personal vehicles that society is imposing on the public infrastructure of roadways. The increasing level of traffic is at the root of harm from motor vehicle collisions but also a lot more. GHG emissions, and energy use, are also proportional to the number of vehicles, the number of trips they make and the length of those trips.

These negative tendencies have so far been blunted by concerted technical actions. Once vehicles have been manufactured to a specified minimum level of safety and powertrain efficiency those vehicles are turned loose in the roads-and-traffic "bubble". This bubble has certain well-established ways of doing things, including many highly professionalized processes and activities. Users need to respect these norms, but users are then able to decide when and where they need to drive in order to discharge all elements of their daily lives. There is a strong momentum to our primary life purposes, mostly freely chosen by each individual, and driving is regarded as being the most essential help-mate – after having a roof over your head. This is evidenced in figures compiled in 2022 by the U.S. Bureau of Labor Statistics (5): the big three categories of household expenditure are housing (33.3%), transportation (16.8%) and food (12.8%).

Limits of time and money, and competing demands on that time and money, determine how much driving actually gets done by each vehicle. Aggregated over a large population, we impose, and superimpose, trips and create "traffic". Currently, there are no artificial limits on trips or traffic. As long as you have a conforming vehicle and driver's license, follow the road rules and pay occasional tolls, there are no general measures designed to discourage or penalize driving for the sake of society.

The conforming vehicle is society's secret weapon. And the norms of road design and traffic engineering have a long-term effect in denying the inevitability of worse things to come. These two elaborate but separate methodologies – the vehicles and the roads – have helped us to largely avoid the need for Garrett Hardin's "mutual coercion, mutually agreed". His concept of the Tragedy of the Commons (6) – the relentless pull to abusive over-use – has been partly averted in the case of the ground transportation system. But the globally observed "tragedy" does not relate to safety; the collective release of carbon emissions from personal vehicles represents 15% of GHG emissions worldwide. Within the United States this figure is approximately doubled, at 29% of all GHG (7). Transportation systems have evolved rapidly in the 21st century through technical interventions but are not able to keep up with widely published disdain for energy use and emissions. We are yet to see evolution play out in the moderation of vehicle usage. Of course this would be doubly beneficial – for vehicle emissions as well as safety.

As we have seen, transportation is far from being a crisp machine. The infrastructure commons and its roads and traffic lie at the center of the transportation economy. This economy exists at the behest of its users, citizens, people, families, drivers and passengers. Hopefully, they do not lead tragic lives of grinding inevitability, to again use Garrett Hardin's term. But they do bring unpredictable and irrational behavior to the commons. On the other hand, the cohort of users make financial sacrifices because someone has to pay for conforming vehicles and functional infrastructure. They also vote with their feet in going along with the prevailing methodologies of transportation. And they contribute more to keeping tragedy at bay by adopting mobility innovations. Transportation innovators are creating new types of services that rely entirely on a closer understanding of user desires for functionality, but such services also need to live and breathe sustainability. Beyond this, the evolution of prolific vehicle users whose minds are on other things may be the final frontier of transportation sustainability.

4.2 SAFETY AS A FUNCTIONAL COMPONENT OF THE TRANSPORTATION SYSTEM

4.2.1 WHICH SUB-SYSTEMS ARE MOST SAFETY-CRITICAL?

It is clear that the "transportation system" is not really a single system, certainly not a closed professional system that could be held to account on matters of principle. That may explain why safety has always worked in reverse: track down the harmful collisions and injuries, understand the scenarios and then cancel them. But which among transportation sub-systems – that make up the complex system of transportation – would be the strongest candidates for creating better, more righteous, systems? And at the same time, which systems may be complicit in deficiencies that cause crashes and injuries?

Road transportation represents a large chunk of our economy and covers movement of both people and goods. It is a very sizable portfolio, with loose ties – difficult to visualize and comprehend. It therefore helps to "follow the money". The private side of people movement includes the industries of vehicles, fuel, insurance, finance, for-hire services, tolls, fees and parking. On the public side, various cost centers – many of which are outsourced – include road design, construction and maintenance, traffic control, education and enforcement. Freight movement covers both own-account and for-hire carriers, trucks and trailers, a significant component of fuel and drivers' wages, depots, terminals, yards, warehousing and rest areas. Freight movement is subject to many kinds of regulations, fees and charges levied by states. Because commercial truckers use the public road system, there is a general political commitment to recovering costs of design and construction to accommodate the unusual weight and size of freight trucks, as well as wear and tear.

Among all of these cost centers *vehicles* have evolved as the most effective flag-bearer for road safety improvement. This started with the severe criticism and regulatory actions of the mid-20th century, moved through protective and assistive automotive designs and has continued with collision avoidance in the 21st century. The current trajectory is automation, complementing a stunning arc of engineering and technology aimed at safety. Vehicles are designed to do more than cancel certain crash types – they have a growing influence on the creation of a safe system, the almost mystical goal of 21st-century road safety strategy.

Another important cost center from the safety perspective is automotive insurance. The high cost of crashes is almost entirely borne by vehicle owners from coast to coast who pay their auto insurance every year. This financial burden is skewed so that the majority who do not crash cover the costs of those who do. But the mechanisms of auto insurance are not sufficiently transparent to comprehend their influence on safety.

The safe system movement started out bringing together the traditional components of vehicles, drivers and infrastructure (or environment). In addition to vehicles, the safe system is advancing mostly through the related components of enforcement, traffic control and road design. The "safe system" strategies adopted in many countries identify road designers as having lagged in their safety contribution. Note that drivers per se are not included in this group; collectively, they cannot be relied upon for safety improvement. On the contrary, they consistently cause crashes through careless driving behavior. Safe system proponents tend to downplay the blaming of drivers and look for other lesser known solutions.

In order to fully understand road safety, we should go back to fundamentals and focus on the various components of the transportation system. We should not limit ourselves to those selected by Vision Zero and considered in the "safe system". The latter terminology is tantalizingly correct, but the branded meaning applies to a philosophy, not a functional system. The components of that functional system will clearly include vehicles, drivers and infrastructure, plus some new elements.

The bright spot of safe vehicle design is now being aided and abetted by Intelligent Transportation Systems (ITS), a new component in its own right; the advent of vehicle and infrastructure connectivity goes a long way to defining a safe system if it can be deployed on a wide enough scale. The much-heralded safety potential of vehicle automation is compelling but misunderstood. It is often described carelessly as if it were another, smarter, vehicle safety technology. However, vehicle automation alters the transportation safe system in a fundamental way: it adds the new component of driving machines alongside the existing components of vehicles, drivers and infrastructure.

Despite the many more components of the total transportation system listed above, the functional components most critical for the evolution of road safety are as follows:

- Vehicles;
- Drivers;
- Vulnerable road users (VRUs);
- Driving machines;
- Infrastructure; and
- ITS.

The road safety system has been traditionally made up of vehicles, drivers and infrastructure. This system is evolving to include VRUs, driving machines and ITS. While driving machines are still in their developmental stages, ITS is a mature but under-utilized component in its own right. VRUs are the most amorphous segment.

Other parts of the transportation system, including traffic control and on-road enforcement, also impact the road safety system but do not represent critical components on a par with the five listed in the previous paragraph. Over a long period, road agencies have been slow to deploy traffic controls through the infrastructure, particularly at intersections where too much is left to the drivers. Enforcement of road rules has been on the rise through greater use of technological means, such as speed cameras and red light cameras. However, the road rules themselves have only a qualitative relationship with collisions and injuries.

4.2.2 GOVERNMENT ROLE IN SYSTEMATIZING TRANSPORTATION

If a number of components are needed to operate a system, how do they inter-relate? It is important to understand the ties that may act to bind a multitude of transportation and safety elements together. The influence of governments – federal, state, city and local – is obviously felt by all players but in different ways. The U.S. Department of Transportation, which did not exist until the mid-20th century, incorporates separate "modal" administrations. The most prominent of these are the National Highway Traffic Safety Administration (NHTSA) and the FHWA, along with the FMCSA and the Federal Transit Administration (FTA). Other federal departments also have an important influence, including the Department of Commerce where the EPA is located.

The supreme oversight agency of the U.S. Congress is the Government Accountability Office (GAO); the GAO maintains a strong focus on actions and plans throughout the U.S. Department of Transportation. At the state level, all have departments of transportation that cover roads and bridges, vehicle registration, traffic, safety and fees and charges. In many states, only a minority of roads are "owned" by the state, and the majority are managed by counties and cities. Driver licensing is carried out by state police and enforcement by locally organized highway patrol departments.

Private companies range all across the road transport industry and make common cause as required, often for lobbying or policy influence related to pending regulation or changes in economic conditions. The vehicle manufacturers and their various industry associations – both light and heavy vehicles – are clearly important players and may constitute regulatory cartels. Their continuous stream of transactions with NHTSA, FMCSA and others represent a closed, professional system with clear principles of conduct and process. This really constitutes two systems: one to handle vehicle safety and one to handle vehicle emissions.

The other major federal player, FHWA, has more of a standards-setting brief than a regulatory role. FHWA deals with the DOTs of all 50 states in setting standards for roads and traffic. The states caucus together in the form of the American Association of State Highway Transportation Officials (AASHTO). This set of activities involving FHWA, AASHTO and DOTs is extensive, continuous and professional and is closed. As a system, it lacks the cause-and-effect urgency of the two vehicle systems.

Organizations like the American Automobile Association (AAA), itself a network of federal and state organizations, represent vehicle owners and drivers. Along with many other industry and professional organizations, the AAA has sufficient status to be consulted on government investigations and rulemaking. These activities are too diffuse to rise to the level of a system.

Despite the broad range of active government agencies in the transportation field, only a few play a major role in system creation. NHTSA plays a critical role in the design of vehicles for safety. The vehicle element of the road safety system is built on regulatory synergy between the industry – who do the real work – and NHTSA. At the same time, the vehicle industry maintains a massive vendor–customer relationship with America's drivers. But it exerts limited influence over driver behavior once they leave the show room. An analogous relationship exists in the field of vehicle emissions control, where the EPA forces continuous improvement in powertrains; EPA's decisions are placed into an industry regulatory framework with the help of NHTSA.

These important aspects of system behavior were forged via the practice of road safety. Safety led the way for all of transportation. The need to regulate occupant protection in motor vehicles created the need for new methods of research, innovation, consultation, legislation, economic assessment and regulation. Relationships between distinct elements of ecosystems are only developed through necessity. The avoidance of serious injuries and fatalities is a powerful form of necessity.

When it came to infrastructure, the same intensity of need did not exist. The state transportation agencies that design and own roadways do not have direct relationships with vehicle drivers or the vehicle industry. These agencies do have relationships with traffic control vendors, but they are one-way and commercial in nature. Therefore, the infrastructure and vehicle parts of the road safety system are historically separate. In no small way, this may be traced back to the mid-20th-century founding of the U.S. Department of Transportation. This was successfully legislated on the basis that the "modal administrations" contained within the DOT would operate separately, with each of the administrators reporting directly to the Secretary of Transportation. It was therefore pre-ordained that NHTSA would stick to vehicle safety regulation and FHWA would concern itself with highway standards, but the twain would not meet, at least within the federal bureaucracy.

4.2.3 NATIONAL SYSTEM OF TRANSPORTATION RESEARCH

There are many other ways in which parts of the transportation portfolio interact, but one of the most valuable is transportation research. The National Academies of Science, Engineering and Medicine include the Transportation Research Board (TRB) which was presciently established in 1920. TRB is mainly focused on the roads and traffic part of the portfolio, rather than vehicles. Among many benefits, it provides the state DOTs with an R&D capability but also identifies major national challenges in transportation and hosts an informed, professional exchange that rises to an academic level. This focal point for research has most of the characteristics of a professional system but is not closed. However, there are clear qualifications for taking part.

4.2.4 THE MECHANICS OF THE PORTFOLIO OF TRANSPORTATION

In an overall sense, transportation stands as a portfolio that contains a number of professional sub-systems, some more specific than others. These sub-systems include automotive, energy, environment, infrastructure, logistics, research, regulations, safety and traffic. Each is built on professional principles and is at least partly open to external influences. The portfolio of transportation generates action and commerce based on a broad set of industries. It exists to serve the consumers, or users, who pay for everything.

Transportation operates via an inter-connected constellation of franchises. These include vehicles, roads, fuel, traffic control, regulators, insurance, charging, mobility, freight, transit and enforcement. The franchises of transportation are very widely distributed so that very similar, but not identical, services are made available in every corner of the country. Collectively, these franchises are required to bear society's weight of expectations. For this, they utilize a range of progressive mechanisms, each relying on freedom of association and information exchange. Such mechanisms that impinge on public consciousness include safety, energy efficiency and traffic efficiency. In addition, social coherence is emerging as an important transportation responsibility expressed collectively through equitable access, mobility upon demand and urban design. State departments of transportation and planning, as well as city administrations, drive these newer activities.

Each of these mechanisms contains several highly professional sub-systems which may have "system administrators" and virtuous overtones. Important roles akin to system administrators are played by two of the U.S. DOT's modal agencies – NHTSA (for safety) and FHWA (for traffic). Located separately in the U.S. Department of Commerce, the EPA also plays a key role for energy and environment. These administrations do not have moral autonomy and their programs are controlled by Congress. The mechanisms for which they play leading roles – safety, energy, environment and traffic – are influenced by the currents of government policy, political discourse and social media, and are informed by professional bodies, academia and technological innovation. As with all matters of government policy and regulation, there are over-riding social imperatives at work; transportation is influenced by its unarticulated interplay of fairness and personal liberty.

4.2.5 Transportation's Earned Virtue – Safety

Does it make sense to consider the philosophies of road safety in isolation from transportation as a whole? Clearly not – safety exists alongside a number of other important values in the portfolio of transportation. Even if transportation lacks a clean taxonomy and a set of universal principles, it is the biggest personal action set embraced by the people of America. And its biggest money pit. It creates endless circulation of people, machines, commerce and culture; in service of people's most important needs and desires.

And it continues, growing and evolving year after year, decade after decade. It is far too active, hungry and pragmatic to stop and understand itself or fix all of its problems. Nevertheless, safety has the distinction of being transportation's conscience and memory; the only one in the family that's been to college. Road safety has a rich, informed functionality but keeps it under wraps so it won't embarrass its siblings.

4.3 THE OVER-WORKED COMMONS OF ROAD TRANSPORTATION

4.3.1 The Tragedy of the Commons

Garrett Hardin (6) was an uncompromising biological scientist who presented a remarkable speech, largely in response to global over-population, at a convocation of the American Association for the Advancement of Science in 1968. The speech was published in Nature in December of that year and was entitled The Tragedy of the Commons. The title is important. In addition to being memorable, it uses the word "tragedy" in its Ancient Greek sense of grinding inevitability. And the word "commons" invokes William Forster Lloyd's 1832 explanation of over-use of the English village commons for grazing cattle. Each herdsman's gain from adding one more animal far exceeds his share of the incremental decline in the condition of the pasture. So everybody does it, and we're all on the highway to hell.

Hardin used this analogy as part of a larger treatise on the fact that some world problems are not susceptible to technical solutions and require changes in human attitudes and abandonment of habitual behavior. There is a need to "legislate temperance". How should the social contract allow for such a mechanism? Hardin's thesis also goes beyond the individual incentive to do the wrong thing to the idea that the impact of doing the wrong thing varies. "The morality of an act is a function of the state of the system at the time it is performed". Pollution of the atmosphere is more harmful in a densely populated area. As for water, Hardin quotes the frontier aphorism that "flowing water purifies itself every ten miles".

Hardin had no illusions that appeals to human decency and sense of responsibility would be effective; he had made a strong point about the vortex of over-use. In his view the playing field of the commons needs to be tilted. In his words, we need "mutual coercion, mutually agreed upon". An effective form of coercion would be financial: taxing or charging. He gave the example of downtown parking and the introduction of parking meters and fines for infringements. With exceptions, punitive charging for vehicular use of the commons of streets and roads has made little progress as a tool of public policy in the United States. But Hardin's thinking has persisted in transportation circles and has been influential in growing opposition to our reliance on cars and in the movement for sustainable transportation.

4.3.2 CYBER SOLUTIONS APPLIED IN TRANSPORTATION

Let us fast forward to 2020 when a Forbes article invoked Hardin to advocate for telecommuting, a useful necessity unearthed during the pandemic disruption of 2020–2021. The Forbes article uses examples of cities given over to parking, badly congested roadways and even gridlock. Since Hardin's striking piece first saw the light of day, the unintended negative consequences of motorization have received extensive attention in professional circles and in the media. These impacts include safety, traffic congestion, pollution, diversion of freight from rail to road, the built environment, urban dislocation, urban sprawl, the global politics of oil and carbon emissions. Along the way, a number of these issues gathered under the academic rubric of sustainability, sometimes known as triple bottom line business practices, and more recently resilience and the circular economy. Some have made observations about the peaking of these matters: "peak oil", "peak car" and "peak driver" – in the latter case, less interest among the younger cohort in obtaining driving licenses.

Global management and engineering consultants have brought forward transformational paradigms, many of which could be bundled under the Fourth Industrial Revolution espousing massive personal use of technology. This helped usher in the worldwide phenomenon of Artificial Intelligence (AI). AI started out as a cybertechnical application but has blossomed to a global phenomenon over a decade or so. Often, the field of transportation has found itself being used as a showcase for commercial cyber paradigms. However, large scale application and commercialization often migrates into other business sectors.

During the first two decades of the 21st century, a number of movements have not only used transportation as an exemplar but are having a significant influence on the practice of transportation. These include Smart Cities, the Internet of Things, Mobile Communications, Big Data, Open Data, Connected and Automated Vehicles (CAV), Corridor Management, Electric Vehicles (EVs), Ride-Hailing, Shared Mobility, Micromobility, Complete Streets, Road Pricing and Congestion Management. Transportation has been a sponge for technology sets that lead to trials, learnings, model deployments and sometimes commercialization. Government cooperation has generally been forthcoming, given the promise of societal benefit, especially in safety, sustainability and social cohesion.

In some cases, the yokes placed on businesses previously inured to government controls have been thrown off. The business model of ride hailing was considered to lie in a grey area crisscrossed with regulation. Would it be stifled by safety regulations? The transportation sector saw innovation win out. An even more interesting example is the commercial use of automated vehicles (AVs) without drivers. Despite endless ways in which safety objections could proliferate and block, AV manufacturers are steadily accumulating the early deployment miles they need.

It is important to remember that these new technologies look for early application in transportation because it is a large and open sector of the economy. Whether we call them technology sets, or movements, or practices, or hot topics, it takes considerable time for them to grow up and evolve. In the 21st century, transportation has shown refreshing verve in changing the prevailing model. It has definitely shown that generalized safety concerns will not stop new mobility paradigms before they even get started.

4.3.3 WORLDWIDE MOVEMENT FOR SUSTAINABILITY

The morality of sustainability has captured the imagination of professionals, academics and citizens alike but has never been strictly defined. Often, it would be no more than "we don't want to leave all these problems for our children and grandchildren". The United Nations (UN) talks about meeting the needs of the present without compromising the ability of future generations to meet their own needs. In business, the "triple bottom line" refers to business reporting of environmental and social results as well as financial results. In financial circles, investors may seek stocks that resonate with goals of Environment, Sustainability and Governance (ESG). Such imperatives have flourished in the new century.

Many academic institutions seem to feel that sustainability is now sufficiently well understood, and it may not need a formal definition. One exception is the University of California Los Angeles (UCLA) Sustainability Committee, who define sustainability as:

> the integration of environmental health, social equity and economic vitality in order to create thriving, healthy, diverse and resilient communities for this generation and generations to come. The practice of sustainability recognizes how these issues are interconnected and requires a systems approach and an acknowledgement of complexity.

While safety is undoubtedly part of sustainability, there has been a tendency for safety to remain separate, perhaps because safety is well endowed from a scientific and professional perspective, and is long recognized among government agencies.

The several tragedies of transportation are well documented and referenced in sustainability circles and in many of the above movements. But the meaning of the commons itself also deserves modern-day consideration. Since the time of Hardin, the concept has been widely applied in many fields, a prime example being the internet. In a 2012 Atlantic article (8), Bill Davidow advocates regulation of the online commons – the internet – because "virtual over-grazing" can occur instantly. He draws attention to the pollution of the internet with many actions that erode personal privacy and can be done at zero cost to the perpetrators. And he calls out this very loss of privacy resulting in an even greater loss of precious personal time dealing with passwords, junk mail and identity theft. He is extremely critical of the tech industry's advocation of self-regulation.

4.3.4 EVOLUTION OF THE INFRASTRUCTURE COMMONS

Economists regard a commons as a good that is "rivalrous and non-excludable". That is, use by one person degrades the simultaneous use by others, and people cannot be excluded. In the case of our system of roads and traffic, the former is true and the latter is mostly true in practice (but not in theory). Forms of exclusion are probably rare enough that our road and traffic system is mostly a commons. In the earlier days of 20th-century motorization, a battle of the commons was fought in towns and cities all over America. Citizens were expected to live with the toll of serious injuries and death caused when fast, aggressive motor vehicles plied the roadways instead of horses and carriages. Many negotiations and changes were required to arrive at a stable, workable system.

Personal discretion to make trips, and sheer speed, propelled the roads and traffic system for a very long time. As such, the infrastructure commons became an absolutely indispensable part of our lives but hardly a beloved part of our lives. Many people are enamored of their motor cars, but once inside they are isolated from the commons of infrastructure and from their fellow citizens who are encased in their own vehicles. Occasional eye contact is the norm of communication. Many people's interaction with the infrastructure at large mostly occurs through the medium of their car.

Our complicated attitudes to the infrastructure commons may be illustrated by the case of the invention of the freeway or motorway. This, the mightiest class of roadways, was called a *limited-access facility*. As Earl Swift explains in his book Big Roads (9), that term did not refer to limiting interruptions to free movement but the limiting of commerce along the right of way. We did not want places of business – such as tattoo parlors and hamburger outlets – appearing along the edge of the roadway. If that occurred, drivers would pull over and cause impediments to traffic flow: the road frontage would become encrusted with blockages and even crashes. The practice of limited access has been continued to this day, and everyone knows you need to exit the freeway in order to obtain comfort and sustenance. The principle of limited access has been validated historically by user acceptance, and motorists taking trips within big cities make extensive use of freeways, rather than take the more diverse and colorful local roads, or surface streets.

Real change comes slowly to a roads and traffic assemblage that is required to host more than 3 trillion personal trips per day. There is no down time. But this system is one of the world's great evolution accelerators, and safety has evolved through many imperceptible changes in roads and traffic behavior. Increasing urbanization has led to more influence of vocal city populations on the form and use of the commons. Less acceptance of the motor car in big cities has brought about physical changes such as those encouraging active and micro modes, including walking and biking but also new, faster electric devices. Every day, traffic lane space is being turned over to these alternative modes. This has brought about an increase in deaths and injuries occurring to those active in these modes, collectively known as VRUs. That is, vulnerable relative to the car. The full range of VRU safety impacts are not accurately known at this point in time, partly because traditional crash and injury statistics tend to be state based. The safety profession has not yet caught up with diverse, city based incidents and data.

4.3.5 SUSTAINABILITY CO-EXISTING WITH SAFETY

The VRU categories that are part of the traditional safety data architecture – pedestrians and cyclists – are showing a very adverse trend. Recent national media attention – including the New York Times and Forbes – has highlighted a 41-year high in pedestrian deaths. Data produced by the Insurance Institute for Highway Safety (IIHS) (10) shows that the contribution of pedestrians to total road deaths has increased from 11% to 17% since the turn of the century. And the annual 21st-century total is as high as it has ever been. According to the Governors' Highway Safety Association (GHSA) (11), most of the modern-day pedestrian fatalities (60.4%) occur on urban arterials that are designed to move large volumes of cars quickly. And many occur where the infrastructure is not designed for pedestrians, where there are no sidewalks.

The infrastructure commons contributes directly to the pleasures of city living. The ancient piazzas, fountains and squares of Europe remain irresistible tourist attractions. Newer forms of outdoor entertainment, conviviality and relaxation are in evidence in cities all over the world. These include outdoor eateries, concerts, street theatre and communal viewing of major sporting events. And city streets have long hosted all manner of parades, gatherings, processions, celebrations, festivals and religious events. Everyone is aware of hot August nights in New York City, with families trying to cool off on the stoop. People everywhere walk their dogs and wash their cars. Attractive and safe places to do all of these things in a totally impromptu manner have irresistible, but intangible, social value. Open and free use for a multitude of purposes is expected. Often-times the city will close off streets to better accommodate such events. And impacts of major events on traffic have become a major issue for city administrations; at the same time, cities want to be judged by the quality and attractiveness of their events.

Generally speaking, the utility of public infrastructure in cities for social purposes, on the one hand, and serious traffic movement, on the other, have been separate domains. With greater urban populations, burgeoning tourism and more large events there is a little more cross-over between the two types of use. Nevertheless, our road and traffic infrastructure does closely resemble Hardin's idea of a Commons. Cars instead of cows. In a sense, access is granted when a vehicle is registered for use on public roads – and, separately, the driver is licensed.

Subject to the registration and licensing mentioned above, and other forms of curation, the transportation commons has remained free and equal to a significant extent. And usage has increased inexorably. According to FHWA data (12), total VMT increased from 1.13 trillion miles in 1971 to 3.26 trillion miles in 2022 – an increase of 188% over that half-century. And this with a smaller increase in the extent of the road system: an approximately 72% increase in paved miles over the same period (1970–2020). This scale of increase in use even exceeds the grinding inevitability implied by the word "tragedy". Perhaps we have seen, and continue to see, the *miracle* of the commons. But it is right to ask how long this miracle can last.

4.3.6 PRIMACY OF CLIMATE CHANGE

While all this grinding has been going on, one aspect of sustainability – climate change – has captured the attention of citizens all over the world and has developed huge political influence. It has long been recognized that GHGs cause cumulative harm to the Earth's atmosphere and that one of the largest man-made sources is the collective exhaust emissions of motor vehicles. According to the U.S. EPA, the economic sector of transportation is the biggest emitter of GHG and currently accounts for 29% of the U.S. total. To put this large number in perspective, the other most carbon-intensive sectors of the economy are electricity generation (31%) and industry (30%).

Over 94% of fuel used in transportation is petroleum-based: mainly gasoline and diesel (13). Cars overwhelmingly favor gasoline, while diesel is dominant for medium and heavy duty trucks and buses. Light-duty vehicles account for 58% of GHG emissions and 23% are caused by the larger diesel vehicles. Energy use in the U.S. vehicle park remains staunchly wedded to gasoline (52%) and distillates (22%) with biofuels (6%) and natural gas (5%) adding to the mix. Electricity for charging EVs barely registers on this scale, but expectations are high.

There is no doubt that the wide use of EVs will greatly reduce GHG emissions. On the basis of total lifecycle, including manufacturing and many years of use, the EPA states that EVs have a GHG footprint less than half that of cars with internal combustion (IC) engines. Complete conversion of the U.S. fleet would therefore make a huge dent in annual GHG emissions but would not go close to eliminating them. Even though IC engines have been improved greatly in recent years, as they have responded to federal rules for fuel efficiency, increases in the number of vehicles and miles driven have won out, and GHG emissions from IC vehicles have continued to increase. Political pressure to reduce GHG emissions therefore signals a serious rollout of EVs – there is no alternative.

According to the U.S. Bureau of Transportation Statistics (14), sales of full battery EVs increased to 460,000 in 2021 but still only represented about 3% of new vehicle sales.

There are some indications that an exponential EV sales curve is being hampered by lack of convenient charging stations and other fears of the unknown. A meaningful reduction in GHG emissions from the transportation sector is therefore a long way off, considering the fact that the 97% of vehicles sold in 2021 with IC engines are going to last until the mid-2030s.

Our infrastructure commons therefore not only plays host to numerous fatalities and serious injuries but also an intractable sacrifice of fossil energy. Are these two separate tragedies? Or are they part of the same pattern of peoples' indifference and selfishness? On the face of it, they are different in terms of the human factor. Road safety is dictated by drivers' every-day, in-the-moment behavior and desire to drive further, more often, faster. Energy use is dictated by past corporate decisions and federal rule-making in the arenas of vehicles, powertrains and fuels. But each has many other layers of influence, and many of these influences are held in common. For example, safety is affected by boardroom decisions and federal rule-making, just like energy use and energy use is affected by driving intensity, just like safety.

4.3.7 COERCION UPON THE COMMONS

Garrett Hardin was in favor of coercion when things are getting out of hand on the commons. He was definitely not in favor of waiting for everyone to do the right thing. From the perspective of safety, drivers have long been subject to some coercion – rules, the enforcement of those rules and significant penalties. While this is somewhat true for vehicle propulsion, powertrain rules are not directly apparent to drivers but are applied to the vehicle and fuel before they get to the point of sale. The abstract and remote nature of these rules may not look like coercion to the vehicle owner or driver. If anyone is coerced it is the vehicle manufacturer, but this is a highly negotiated and stylized form of coercion. Hardin's idea of coercion was mutual: "we're all in this together and we'll all agree to accept a little pain for the common cause". This concept probably exists in safety, where road rules apply equally to all, but not in vehicle emissions.

While the need for safety rules is readily accepted within the infrastructure commons, the pressure for energy rules – especially related to climate change – comes from without. Concern about climate change is global and impacts every sector of national economies and world commerce. Even though transportation is the biggest generator of GHG, the commons is surprisingly free of restrictions. This is starting to change, with a few cities declaring bans on IC engines and states contemplating similar moves. The external pressure to reduce GHG will continue and intensify, and place-based restrictions on powertrains may begin to have a direct impact on car owners.

4.3.8 FORCED REDUCTIONS IN TRAFFIC CONGESTION

Traffic congestion is a very clear case of the tragedy of the commons. Like the harm caused by crashes, it is a vice endemic to the infrastructure commons. Because vehicle ownership represents one of the biggest investments made by households, it is rational for vehicles to be driven more and more miles. This makes sense because the marginal cost of driving more miles is tiny relative to the up-front cost of owning a vehicle. But then everyone suffers the consequences of delays, stoppages and traffic snarls.

Currently, attempts to limit the volume of cars in cities, or parts of cities, are the exception rather than the rule. Various forms of pricing may be used to restrict trips into the city center. Such schemes are unpopular unless you happen to live in the center. And experience in a number of cities shows the need to increase charges from time to time in order to maintain the goal of deterrence. The city state of Singapore is well known for firm restrictions on cars entering the city. This includes a high-priced license to own a car in the first place, a prohibitive sales tax and the usage-based charge itself. This is financial coercion writ large! Such measures will not be tolerated by residents unless there is a high-quality and reasonably priced alternative: this means excellence in public transport. There are therefore very few Singapore's around the world, with London probably the closest.

4.3.9 THE VICES CAUSING OVER-USE

The national infrastructure commons is being sorely tested by complex vices arising from the way it is used. Its duty cycle is derivative – the net result of numerous primary demands, including employment, education and entertainment. Those demands serve the most important needs of the vast majority of our society; they are irresistible and sometimes all-consuming. The commons is there to be exploited and there is no proprietorial constituency for moderation, much less ethical use.

The moral blight of fatalities and injuries is monitored, quantified and countered in a highly professional manner. But the science of road safety is almost entirely utilitarian. And it is the province of government transportation agencies, not individual vehicle owners and drivers. There is little room for reflection on moral paths or wise maxims to be followed. Vehicle owners and drivers are constantly reminded of their duty of care to drive safely. However, when they are driving home from work or a night out, or trying to satisfy the desires of the entire family on a trip to the mall, or catching up on phone calls, they are only incidentally focused on the driving task. And their perception of crashes as such rare events means that the creeping incrementalism of Hardin does not enter their calculations. They do not view the sheer volume of other road users as a threat to the well-being of themselves or their families. While they may look for a five-star safety rating on their vehicle purchase, there is no real moral dimension to their driving. Never a thought to reducing their mileage – and therefore exposure to collisions – in order to be safer.

In the case of GHG emissions, a massive worldwide cohort who espouse *technomorals* has spoken with a single voice: a virtuous society must eliminate climate change. This may be expressed through carbon neutrality – taking as much carbon from the air as you put in – or net-zero carbon, requiring a reduction in your carbon as well as taking some carbon out. These maxims are combined with long-term utilitarian targets for global temperature increases. This climate application of virtue ethics and deontology applies to public and private institutions but tends not to touch consumers directly. While everyone is aware of electric vehicles (EVs), there is no compulsion to buy one. And nobody is telling you, en masse, to reduce your IC vehicle mileage in order to reduce your GHG output.

Traffic congestion is the most visible failing of the commons. But it lacks the moral weight of lives lost and the repudiation of GHG emissions. Most people's solution to backed-up traffic is more and better infrastructure. It is obvious to everyone that the unheralded addition of more vehicles to the commons is causing more hold-ups, but individuals do not feel any sense of ownership of the problem, nor responsibility for solving it. State and local road authorities may receive reputational damage from recurring congestion, but their responses are always limited by state budgets. And citizens tend to over-estimate the costliness of infrastructure and are not looking for more expenditure. Expectations are therefore low. So road authorities focus more and more on improved operations – getting more throughput on existing infrastructure – but gravity eventually takes over.

4.3.10   THE DYNAMICS OF CONGESTION

Congestion occurs because more and more cars on a given roadway are compressed closer together until a few drivers feel less comfortable about driving at the free speed. When they ease off a little and slow down, everyone has to slow down. And because drivers are imperfect, each slows down fractionally more than they need to. Pretty soon the whole roadway is operating at a much lower speed. Or may even experience a "shock wave" and become stationary. This is the behavior of a cohort of vehicles traveling down or through a piece of road. It is a dynamic version – a perfect microcosm – of Hardin's tragedy. And the worst drivers always dominate – there is no averaging in a cohort of vehicles following each other in a lane.

Traffic engineers deal mainly with the volume of traffic flow, the density and the speed. Because there is a unique human driver controlling each vehicle, traffic does not follow laws of physics. Measurements and observations are made, and certain patterns repeat and engineers have created empirical descriptions of traffic behavior – the *fundamental diagrams*. This working knowledge applies at the macroscopic level and is oblivious to the actions of individual drivers. The first fundamental diagram known by all traffic engineers tells us that the volume of traffic increases with its density up to a certain point and then starts to reduce. If we want to have the greatest number of vehicles per hour traverse the road, then a certain amount of headway needs to be maintained between each vehicle. If not, the speed will drop and the volume will drop as well. This is pure Hardin and the inevitability is like night following day.

But there is a different point of view that makes sense to traffic engineers. First and foremost, they care about volume, measured in the number of vehicles passing a fixed point per hour – the greatest utility from a piece of infrastructure. But the user cares about speed because that determines travel time and in the negative creates delay and uncertainty in arrival time. Another fundamental diagram tells us that speed reduces with increasing density, and this starts to occur well before maximum volume occurs. So the introduction of one more vehicle, and another, and so on, is a serious detriment to the user of the road well before it thwarts the best laid plans of the road agency. But the road agency is the custodian of the road network, and their strong work ethic as seekers of maximum volume does not quite align with users' desire for maximum mobility. Users are probably not aware of this, nor of traffic volumes in general. So the agency's ethic of more volume, expressed as greatest use of the infrastructure investment, prevails. The tragedy of the commons, as perpetrated on the user, therefore helps the agency obtain more traffic volume. Then a certain point is reached where all bets are off, and both speed and volume collapse, and nobody is happy.

4.3.11 TRAFFIC'S EFFECT ON SAFETY AND SUSTAINABILITY

The mobility-enhancing properties of traffic, including the avoidance of congestion and delays, represent an important set of values in the operation of the commons. They are often treated in isolation from safety and sustainability values. Perhaps because traffic is engineered using empirical diagrams, rather than laws, little is known of the effect of road traffic variables – volume and density – on safety. In a general sense it is considered that lower speed is safer, but volume and density do not translate well into traffic conflicts or the risks of collisions occurring in a stream of traffic.

Turning to the case of sustainability – GHG emissions in particular – the road agency's goal of maximum traffic volume means increases in GHG from motor vehicles. The user's driving behavior has a major effect, not only in the travel speed they would like to adopt but also in the way they get to that speed, and then handle forced speed variations. It is certainly true that low speeds (10–15 mph or less) greatly increase emissions of GHG from traffic.

Several different constituencies are always hard at work on the infrastructure commons, and they each have their own means of operation, set of franchises, values and goals. Safety and traffic mobility are both specific to the commons. Safety has a reputation for scientific rigor and well-earned moral standing, and its constituency is granted considerable independence. Traffic mobility well fits the mold of a Hardin tragedy. By necessity, it is much more vocational than rigorous, is not in a position to be righteous, and its constituency is subject to some confounding influences. Sustainability, in the form of GHG limitation and climate-friendliness, is a well-recognized global virtue that is currently under-played on the commons.

4.3.12 SIDE-STEPPING HARDIN'S TRAGEDY

The infrastructure commons is indeed over-worked, but this is not a major threat to road safety because collision risks have continued to reduce. The tiny incremental risk attached to each additional car on the road network is far outweighed, over time, by improved vehicle performance, driver behavior and road protection. The sheer volume of driving is the main enemy of road safety; fortunately our system of roads and traffic is able to absorb a massive amount of driving with a surprisingly moderate effect on safety. In terms of Hardin's analogy, safety does not deteriorate with more cars on the road because the enterprise of road safety prevents it. And we do not resort to coercion; rather we have developed scientific countermeasures. In Hardin's terms, technical solutions have succeeded and hard coercion was not needed – the opposite of Hardin's contention. For safety, we have not found the need to re-educate people and force broad changes in their motivations and behaviors.

Vehicle energy use and emissions represent a different case but also fails to follow Hardin's formulation. Here, the impacts of over-use stand out as being harmful. Even then, technical solutions are going to be effective in reducing fossil fuel use and carbon emissions. But technology will not offer a complete solution. There will be a need for significant moderation in the volume of driving, even in a widespread minimum-emissions technical scenario.

Finally, the most visible form of deterioration on the infrastructure commons is traffic congestion. In order to accommodate the addition of more cars in the system, the roadway must eventually deal with a higher volume. Even though this is best achieved through higher density and moderated (lower) travel speed, users don't like this mode of traffic. When drivers aren't happy, they slow down – but in a traffic stream everybody then has to slow down. This problem is definitely not amenable to coercion but is open to technical solutions. Such solutions – the technologies of ITS – are seriously lagging in application by highway departments. But perhaps the problem has not yet become sufficiently acute, as occurred for the mid-century epidemic of serious injuries and fatalities.

4.4 THE STEADY HUM OF ROADS AND TRAFFIC

Road transportation operates all day, every day, all over the country. If you emerge from the Portland International Airport, rent a car and drive to a meeting at Freightliner your experience will be similar to arriving at Austin-Bergstrom International Airport outside Austin, TX and driving downtown to the Texas State Capitol. On the surface these are completely different driving environments, but in terms of the credentials, skills and knowledge required, they are basically the same.

Somehow, your Michigan Driver's License is acceptable in Oregon and in Texas. When you get to the car rental desk, Hertz recognizes your Gold Card in both places. Your insurance also follows you around. Your Waze app guides you to your destination using Google Maps. The roads you drive on are not the same, but they follow a familiar hierarchy with freeways, service roads, on-ramps, off-ramps, arterial roads, local surface streets and city streets. When you are sitting at red lights, you may notice that the traffic control cabinets are made by Eagle in Oregon and by Econolite in Austin but you still sit at red lights in both places.

But you also notice that the traffic speeds are higher around Austin, the traffic throughput is beefier, and the sheer scale of the infrastructure is bigger. And you do stop at red lights and watch your speed because traffic police are the same everywhere, or seemingly so. You can also count on your vehicle having a similar set of controls and safety features like airbags and brakes that won't skid on slick patches.

4.4.1 Traffic Professionalism

All of these experiences are the result of the many franchises that operate all over the country, with core similarities and local twists. The engineering principles behind the road system are the responsibility of state DOTs who caucus together on technical matters (AASHTO) and receive assistance from the FHWA. Road design is carried out by large national engineering firms. Traffic control is localized and carried out by county and city administrations, as well as the state DOT. Local officials and engineers call on specialized consultancies, often national, to help them figure out their needs for traffic sensing and control, and ITS. This would include traffic control centers. A relatively small number of companies design and manufacture traffic controls and ITS, with their own layers of suppliers.

Certain cities or counties may take the initiative to install corridors with increased intelligence, either to relieve traffic congestion or to reduce crash rates. Such corridors utilize adaptive signals, and signal coordination between intersections, based on an improved picture of traffic in space and time. Integrations like this, if deemed successful, may be repeated in other locations. The traffic control vendor would be very amenable to repeating the integration elsewhere. So-called variable message signs (VMSs) may be used to tell drivers more about traffic conditions.

In the traffic control center, ever-vigilant operatives are looking for anomalies in key traffic attributes. These include average trip time between two points on the grid, average speeds on certain sections of certain roads ("links") and queue length at intersections. The latter implies the number of stationary vehicles being impeded by a red light. Traffic engineers tend to be conscious of the number of signal cycles required to get stationary vehicles moving. One or two cycles may be OK, but many aficionados would draw the line at three. At a given intersection, it is possible to adjust the timing of the red and green signal phases to take account of worsening traffic efficiency – often called the "level of service" – experienced by drivers attempting to traverse the area.

Traffic congestion may be "recurring" or "non-recurring". For example, traffic volumes vary throughout the day in a somewhat predictable way, as influenced by commuting patterns, school trips and the like. This would also occur over a longer time scale, under the influence of seasons, including weather, vacation periods and economic disruptions. Alternatively, special events – out of the every-day – increasingly cause traffic congestion. These include large spectacles of sports and entertainment. Such traffic is not business as usual and traffic operatives implement special measures that likely involve re-allocation of traffic over a wide area. So-called traffic incidents arise from broken-down vehicles and crashes – this requires "incident management" on the part of the road agency. Even relatively minor crashes cause delays, and severe crashes result in lengthy disruptions because police investigators need to do their work while evidence is still present on the roadway.

The conglomerate of roads and traffic obviously does not run on rails. There are the local variations on a general theme, the unpredictable mix of demands from people going about their daily lives, totally uncoordinated events, changes in atmospheric conditions, breakdowns, lapses and misunderstandings, people changing their mind etc. But like the stock market, all of the pluses and minuses, successes and failures, striving and neglect aggregate to create a recognizable surface and some credible performance indices. These indices, including total VMT and annual average daily travel (AADT) are viewed in retrospect. VMT is available on regional, state and national scales, and may be segmented by road class. AADT speaks to traffic volume on a particular roadway or route.

Like the stock market, we would like to predict the quantitative supply of road traffic, but we have learned to look at it in the rear view mirror. In any case, no one knows how to interpret it as a figure of merit. In fact, it does not stand alone but is the eternal denominator. It represents all the movement we must have. Taking that into account, performance of the "system" is expressed per unit VMT. The most common metric is crashes or lives lost per unit VMT. But road transportation is not really a system – it is too open and complex. Road traffic is more of a system because it is shaped by the design of the roads and the vehicles, and by the rules that apply to both. The resulting collection of professional designs and rules shapes the behavior of drivers but does not control that behavior, or even influence it in a predictable way.

Road traffic does not re-invent itself or start from scratch every day. It operates as if using a professional exemplar as its starting point and constant reminder The exemplar would like to think that it can orchestrate roads and traffic and bring about safety, efficiency and climate neutrality in its many end results and intertwined outcomes. However, the exemplar is not good enough to reliably produce respectable outcomes and may allow harmful outcomes to occur.

4.4.2 The Question of Over-Use

The commons definitely seems to be over-worked from a traffic perspective, and the efficiency of traffic movement is often questioned. Almost certainly the degree of over-work would vary from place to place, but sections of idle infrastructure are almost non-existent. When it comes to measurable trends in nationwide traffic delays, the Texas Transportation Institute (TTI) 2021 Urban Mobility Report (15) represents the state of the art. TTI break down the numbers by size of urban area, class of roadway, hour of day, day of the week and city, covering 494 urban areas. They find that the longest delays occur in the larger cities, with a population of 3 million citizens or more.

One of their two key measures of delay is "Yearly Delay per Auto Commuter", which they determine as the additional travel time, during peak times, when the speed of travel is forced below the free speed. The second measure is Travel Time Index, which is the ratio of the travel time during the peak period to travel time under free-flow conditions. Virtually all TTI's metrics of delay are referenced to free-flow conditions. Significant delay is occurring under off-peak as well as peak conditions, and on streets as well as freeways, as shown in the following 2019 figures for urban areas over 1 million population. Here, we see the percentage increase in travel time, as compared to the travel time required at the free speed:

- Peak freeway   +38%
- Peak street   +30%
- Off-peak freeway   +15%
- Off-peak street   +17%

From a user perspective, traffic efficiency would be sub-standard because travel time is generally greater than that available under free-flow. But there is a good reason for that: speed reduces as soon as you start increasing the density of cars on the roadway, as well as the all-important volume of vehicle movement. The road agency is obliged to seek a different type of efficiency: the volume of cars being accommodated. This will be a maximum when the road looks a lot more crowded, and the available speed is down on the maximum – similar to the off-peak delays shown above – but still very serviceable.

Hours of delay have more than doubled since the 1980s; the anomalous pandemic year of 2020 saw annual delay per commuter drop right back to the levels of the 1980s. Meanwhile, VMT increased by more than 60% over the same period – but became more urbanized – while the population increased by about one third. So the users' view of the state of roads and traffic, based on travel delay, has steadily deteriorated over several decades. This is probably telling us that the professional method of stabilizing traffic movement is indeed having a long-term influence. But it is not sufficient to prevent increases in travel delay subject to mobility market forces. And at least part of the user delay is being caused by road agencies legitimately operating more roads at higher volumes.

Modern road and traffic agencies see their role as operational, not developmental or expansive. It makes perfect sense to aspire to more efficient utilization of the infrastructure assets we have. But the roads and traffic net is not a system with controls and a dashboard. It is more like a market. The professionalized method of traffic on the commons has evolved over many years and consists of designs – of roads, vehicles and traffic controls – and rules that apply to all. That accumulation of techniques, or method, brings about continuity and consistency, so that vocational tools and policies remain relevant year after year. And users retain a sense of comfort and competency that will last a lifetime without abrupt changes. Within reason, the same things happen day after day – and that's a good thing.

4.5 CONFLICTED TRAFFIC IMPACTING SAFETY

The road and traffic system anchors the muscular reality of transportation and has a broad influence on safety, along with vehicle design, which has a more specific influence on safety. The road system is designed for good order, with carefully dimensioned and marked lanes, entry ramps, merges and intersections. The overall geometry is conducive to good behavior, with generous sight lines, grades and curves. Signage and traffic controls are designed to interface well with human drivers, to be realistic and clear, and not too demanding with their instructions and guidance. Designs leave room for a wide variance in the capabilities of drivers and err on the side of driver ignorance and incompetence. There is also a certain margin for error, for example, with emergency lanes on most freeways.

Unlike the legalistic regulation of motor vehicles, regulation of roads and traffic is a mix of standards, guidance and some laws. The regulation of roads and traffic is not mandatory, while laws pertaining to vehicles and drivers are vigorously enforced. But what exactly are the offerings of the roads and traffic franchises? The rate of construction of new roads has slowed over the years, in favor of trying to increase the capacity of existing roadways. These days, state roads budgets are mainly given over to maintenance of the extensive asset and to day-to-day operations in the face of increasing usage. Conditions attached to federal roads and traffic grants play an important part behind the scenes. Some categories of expenditure, such as ITS, are eligible for new construction but not for maintenance. While all of this work is championed and directed by the government agencies, it is largely carried out by a raft of private companies who curate the many requisite competencies.

4.5.1 THE TRAFFIC FRANCHISE

The traffic franchise is more technological than the road construction franchise and more front-line political when things go wrong. Government agencies retain a certain level of expertise prior to outsourcing, and this expertise tends to be located closer to the grassroots. Cities and counties like to stay involved. Unlike the deterministic engineering of a bridge, traffic engineering has to respond to changing conditions and is forced into trial-and-error solutions, supported with real-time monitoring. Traffic engineers have long been focused on efficiency in simple terms of traffic volume, travel time or average speed in traffic. In recent years, there has been more consideration of the users' perspective, and the term "reliability" is used to describe how readily a motorist can predict their time of arrival as a destination. From the institutional perspective, newsworthy aspects of traffic chaos are to be avoided. This would include jams, diversions, queues at freeway on-ramps and off-ramps and all manifestations of clogged intersections.

The day-to-day operations of the roads depend on their quality, reliability and being ready for traffic. For example, large potholes may cause sudden evasive actions and increase crash risk. Trafficability may depend on the size and weight of the vehicles involved and the imposition of special limits. Given the long design life of roadways, and regular surveys of the whole network, the physical aspects of roads are not generally an issue in real time. However, the ever-increasing incidence and scale of roadworks has entangled consequences for safety as well as traffic delay. Road agencies are more and more concerned with ways to reduce the traffic impact of work sites and to prevent crashes between members of the traffic stream and construction vehicles operating at the site. Incident management teams are trained and ready to deal quickly with many different forms of disturbance to traffic. Increasingly, collisions are viewed as unwanted disturbances in the wholesale movement of traffic.

Traffic is market-driven and cannot be regulated. It is an outcome of all the ways that movement in vehicles supports the daily needs and interests of citizens. Certain patterns in traffic volume and density emerge but are generally not predictable. For example, traffic shows hourly patterns that differ by day of the week. These patterns continuously recur, overlaid by frequent incidents and disturbances. Down in the traffic control bunker, experienced and highly attuned operatives are on the lookout for deviations from the norm. They are particularly motivated to anticipate traffic jams and take corrective action.

The connection between these traffic operatives and the humans driving the vehicles that collectively cause disturbance is tenuous to say the least. The operatives can change a limited number of settings – the phase timing of a set of signals, for example – but drivers in the traffic stream may not even be aware of the change. If the signals are programmed to adapt to conditions automatically, few drivers would be aware of this. In a nutshell, the roads, vehicles and drivers are all prepared – ahead of time, each in their own way – to be traffic-ready. It is then up to traffic management to keep order in the moment but not to make noticeable changes.

4.5.2 BAD TRAFFIC

So what constitutes bad traffic, and who would be responsible for fixing it? For a given class of roadway and design speed, the FHWA guidance tells you what the vehicle headway should be, and then the design volume can be calculated. In reality, as the volume increases the speed drops and the travel time increases. Most safety experts are clear that less speed is safer, but this may not be true when the headways – the distance between the front of my vehicle and the rear of yours – are shortened. This is one of the mysteries of traffic behavior that researchers have not yet solved.

However, we do know that safety is compromised when there is a localized mix of stationary and moving vehicles in the same traffic lane. A moving vehicle crashing into the rear of a vehicle that has stopped at a red light is a very common and dangerous type of multi-vehicle collision. Still, it is rare for safety literature to call out unsafe traffic conditions per se. Furthermore, most crash investigation and countermeasure development is limited to the main protagonist – "Vehicle A" – who is considered blameworthy by police and crash investigators. Even though the majority of serious crashes involve more than one vehicle, very little is known about the roles of Vehicle B and Vehicle C or other vehicles that may have been present but did not suffer actual impact. These birds have flown. In most crash databases, the unsafe tendencies in traffic behavior may only be surmised through data elements such as road standard, location, speed zone, traffic control and so on. There is no information about perturbations in the traffic stream.

Bad traffic is more likely to be blamed for violating the environment and more recently the climate. Or for wasting fuel. This would be especially true for freight vehicles, such as Class 8 tractor-trailers operating at 80,000 lb. The constant stopping and starting of these diesel-powered machines as they go about their urban cycle of work consumes fuel and creates a hostile layer of emissions in the form of chemical gases and tiny particles. But again, the currently available solution to the problem lies upstream at the EPA, in the efficiency and cleanliness of engines, rather than in the immediate world of the traffic. Early-stage efforts are being made to create "green wave" corridors that seek to avoid vehicles having to stop as frequently. And smart traffic signals are able to offer signal priority to certain classes of vehicle, such as heavy trucks and buses, but these are relatively little utilized at the present time.

4.5.3 CONFLICTS AT INTERSECTIONS

Roadway intersections clearly create traffic conflict, and it is the job of traffic engineers to manage this conflict. Relatively early in the era of modern road safety, it was recognized that safety analysis of traffic at intersections was far too complex to rely on actual collision frequencies alone – there are not enough of them. The thinking was that analysis of conflicting vehicle movements, and how close vehicles were to colliding, would provide a much faster technique. In the early days, human observers were used to study such conflicts and to rate their severity. However, this complex and time-consuming activity struggled to find wide application, especially because it was subjective and there was no definitive evidence linking conflicts to crashes. Some felt that a crash was simply a conflict that failed to be corrected, while others questioned why the two should be treated as coming from the same species of behavior.

However, new technology including video analytics eventually enabled traffic conflicts to be measured, resulting in metrics such as Time to Collision (TTC). A traffic conflict was defined as "an observable event which could end in an accident unless one of the involved parties slows down, changes lanes or accelerates to avoid collision". The advent of naturalistic driving data further encouraged the use of conflict measures. Such studies showed that the factors affecting real crashes, and those affecting near-crashes, are virtually identical in mechanisms and frequencies. The Virginia Tech 100 Car Naturalistic Study (16) showed that near-crashes occur 10–15 times more often than real crashes. In the case of an impending rear-end collision with a stationary vehicle, heavy emergency braking and/or sudden swerving occur 10–15 times more often than actual collision. This finding, one of the glories of the naturalistic methodology, proves that even a highly conflicted traffic situation allows plenty of scope for crash avoidance. This raises the question: just how bad is bad traffic, from a safety perspective?

4.5.4 THE CASE OF REAR-END COLLISIONS

While traffic conditions are not considered directly in crash investigation, one crash type is often associated with aspects of the traffic stream. The propensity for a rear-end crash in a moving stream of traffic – aside from intersection crashes – is affected by the headways or longitudinal gaps between vehicles. If one vehicle slows dramatically or stops suddenly, will the following vehicles be able to avoid rear-ending the precipitating vehicle? In simple terms, the following driver has a certain reaction time; to this must be added the time needed for their vehicle to slow down sufficiently. This will be a function of the braking performance of the vehicle: deceleration, stopping distance etc. How seriously does this mechanism play into crash causation?

NHTSA's comprehensive 2007 study of pre-crash scenarios (17) tells us that a significant proportion of serious two-vehicle collisions are rear-enders that occur in moderate speed zones. Rear-enders involving more than two vehicles are proportionally more common, including those occurring away from intersections. These findings show that "traffic" in the form of a moving stream plays host to many serious crashes and fatalities. It is interesting that NHTSA did not identify traffic as a causal factor, even though this is normal, traditional safety practice. There is little doubt that traffic speed and density play an important role in such crashes, but neither measure will be found in current crash databases because it cannot currently be measured.

4.5.5 IMPACT OF URBAN RIDE SERVICES

Mobility of people and goods is continuously addressed by entrepreneurs offering new services that save time or money for users. On-demand ride services like Uber and Lyft created a new business model that occupied a grey area in terms of traffic rules. Such services, widely offered as a transportation option in cities, are variously called transport network companies (TNCs) and ride-hailing companies. These emerging franchises, some of which entail the sustainable concept of shared mobility, have an impact on both safety and traffic congestion. The popularity of these services, especially in big cities, shows that convenience and affordability outweigh any concerns about the skills of unlicensed vocational drivers. This represents further evidence that most people happily rely on the remoteness of crash risks.

The available research on these newer services is not clear about their impact on traffic congestion, but some results show a lot of "deadheading" miles being racked up. This would suggest an increase in traffic congestion, but there is scant evidence of such an effect. On the other hand, there is little evidence that TNCs have a positive impact on traffic congestion or much effect on reducing private car usage in cities. They may well detract from traditional transit services like city buses. But some studies show a surprising reduction – up to 25% – in urban fatalities and injuries following the introduction of Uber and Lyft in major cities. There is, however, some evidence of unintended consequences, given the need for users to enter and exit stationary TNC vehicles in convenient downtown locations, where curb space may not be available. This can lead to users being struck by bypassing vehicles. All things considered, the question of the impact of TNCs on urban traffic and safety remains open and a matter for debate.

4.5.6 THE MISSING LINK

Traffic seems to be "the ghost in the machine" when it comes to understanding road safety. The separation of responsibilities in transportation – covering roads and traffic, vehicles, drivers, regulators, enforcement, ride services, insurance and freight – means that safety cannot be neatly corralled to any one of these sectors. Some of these system components – like roads, vehicles and regulation – operate peacefully and quietly, well upstream of the operational moment. Others, like traffic and enforcement, are more "in the moment". In order to understand road safety as an operational reality, we should get as close to collision events, in time and space, as we can. This has been impossible in the past, but technology like ITS is bringing new possibilities. And vehicle automation is forcing a language of "safe passage", a previously intractable concept of each vehicle's buffer zone in motion.

We could view traffic as the sum total of proximate drivers' behavior, subject to the physics of the roadway and the clutch of vehicle capabilities involved. These drivers will vary greatly in their driving performance; and errors and derelictions will be occurring in real time. As we have seen, many other states, situations and actions will conspire to cause an individual crash. These factors will include those remote from the collision as well as those in influential and immediate vicinities. But are these not perturbations applied to the prevailing pre-crash moment? We do not currently have a way of describing that moment, but AV developers may be leading the way with depictions of each vehicle's surrounding "safety zone". Increasing traffic levels and greater vehicle crowding in traffic lanes do affect safety and may be changing drivers' perception of the safety zone needed around their vehicle.

4.6 DETECTION, MONITORING AND SURVEILLANCE

4.6.1 EXPERIENCE WITH SPEED CAMERAS

Vehicle owners and drivers underwrite much of the on-going cost of the transportation system and many aspects of road safety. It is therefore unsurprising that authorities are reluctant to deter drivers from driving or subject them to unduly harsh punishment for transgressions. Nevertheless, the Australian State of Victoria has been a front-runner in driver safety interventions for many years. Milestones included compulsory seat belts (1970), random breath testing (1976) and automatic speed cameras (1989). Speed cameras pioneered automatic enforcement of driving infringements, offering a higher level of detecting driver behavior. They produced very positive results. Cameron (1993) reported a significant reduction of 30% in casualty crashes on arterial roads in Melbourne. Speed cameras have since become a staple of road safety enforcement throughout Australia.

Along the way, point-to-point cameras have been introduced in order to monitor average vehicle speeds on selected sections of roadway. According to a research analysis carried out with Swedish deployments of both speed cameras and "section control", the latter is more effective, reducing all crashes by 20% and fatal crashes by 56%. These studies also confirmed that actual vehicle speeds reduced a similar amount; and the cameras did not have obvious unintended consequences, such as speeding on alternative routes.

4.6.2 EMERGENCE OF ITS FOR SAFETY

There have been many, many studies of speed cameras, including their accuracy and reliability, legal aspects and public opinion. While their efficacy in reducing crashes is widely supported, an element of resistance persists and is very strong in certain societies and locations. We are perhaps finding that surveillance technologies are going to diversify and appear in new morphologies that appease critics. As ITS technology has advanced and received wider deployment, devices for roadside communication with vehicles have become broader in scope and may serve many purposes. Differing bundles of technology may be used on different types of roadways, including highways, arterials and local streets.

Alongside the overall success of well-managed speed cameras, red light cameras have been deployed to a lesser extent and with more controversy. The task of the red light camera is more complex than that of the speed camera, and the possibilities for unintended consequences are increased. For example, a number of studies included in an international review (18), published in 2003, found that side impacts reduced, but rear-end crashes may increase. This review found that traffic signal violations reduced by anywhere from 22% to 78%, and there was also a large "halo effect" at nearby intersections without cameras. When it came to reductions in injuries, the considered finding of this review was that red light cameras reduce injury crashes by 25%–30%.

Increasing use of ITS technologies with multiple purposes means that the extremely valuable safety benefits of speed cameras and red light cameras become blurred. Red light cameras may be incorporated in traffic signals, adding more variables to cost–benefit analyses needed by public agencies. And roadside systems on highways may include tolling as well as safety functions. These combined technologies also raise the stakes when it comes to objections; increased complexity and wider functionality may look more like surveillance and raise the specter of invasion of privacy.

While ITS technologies have vast safety potential, governments have been wary of their surveillance connotations, as well as potential use for collecting road user charges. As road-based sensors moved to adoption of video and radar, new applications opened up in roadway incident management and in construction zones. Limited uses also appeared in tolling for road use, variable speed limits and lane-level traffic measurement, as well as special applications such as wrong-way detection on freeway exit ramps. Despite the unrealized potential of ITS, road agencies worry about deploying soon-to-be-obsolete devices and electronic systems that so obviously lack the gravitas of concrete.

4.6.3 Vehicle Telematics

These infrastructure developments coincided with increased use of vehicle telematics. One of the main drivers for telematics was GPS and the popularity of in-vehicle navigation. Many of these systems were proprietary to vehicle manufacturers. External information flowed to the vehicle in order to give the driver current information including the prevailing speed limit, weather and traffic, in addition to navigation. Information also flowed from the vehicle to a telematics provider – perhaps part of an automotive company, like GM's OnStar – and could be used for tracking vehicles and updating maintenance requirements. Telematics data flows were dynamic but not necessarily real-time.

As personal devices took over many of these functions for private motorists, telematics providers marketed their services for fleet management purposes, more so than individual consumers. Sensing and recording devices were fitted to fleet vehicles in order to monitor driver performance and develop corrective procedures for riskier drivers and for driving practices deemed to be unsafe. Telematics services were widely adopted by fleets of large and heavy freight trucks, especially to monitor hours of service and speeds. These services include location and route, including "geofencing" for vehicles designated to use restricted sets of roadways. At the same time, automotive insurance companies developed Usage Based Insurance (UBI) where a motorist could opt to fit a black box that would generate metrics related to driving style, speeding and aggressive braking. Insurance premiums could be reduced as a reward for safer driving.

These types of information flow to and from vehicles all had pluses and minuses from the perspectives of vehicle owners and drivers, and their use has been voluntary. Telematics has generally fallen short of being a pervasive improver of road safety. Such services have been difficult to value for private motorists and fleet operators alike. And aversion to "big brother" motivations of governments, real or not, is always just below the surface. At the same time, the cause-and-effect connection between monitoring-and-enforcement technology and the prevention of injuries can be tenuous.

4.7 HIGH-FUNCTIONING V2V AND V2I

The new century's developments in vehicle and infrastructure connectivity, centered on high-functioning V2V and V2I communication, have raised the stakes for intelligence that will have a direct impact on crash avoidance. But this new system intelligence needs to be "always on" and cannot be selected by individual vehicle owners or fleets. It is up to national governments to bring about the paradigm of V2X as a functioning risk-averse system, completely endemic to driving.

V2X connectivity fundamentally differs from previous ITS methods of communication with vehicles. It was designed to use affordable bandwidth licensed in the United States for purposes of public safety. It used standardized Dedicated Short Range Communication (DSRC). DSRC radios could be placed in any vehicle or in any part of the road infrastructure. Message formats were created by the U.S. DOT, including the Basic Safety Message (BSM) that streamed the instantaneous position, speed and direction of a vehicle equipped with a DSRC radio. Vehicles traveling in the immediate vicinity, and fitted with the standard equipment, could then calculate a continuous assessment of the risk of physical contact with another vehicle. This system was designed specifically for safety. Once deployed, a large number of "use cases" developed by NHTSA could be used to warn the driver of many and varied types of crashes that may be about to arise.

This approach represents a new generation of in-vehicle Advanced Driver Assistance Systems (ADAS) warning technologies that are more flexible and numerous and much cheaper given that they are not proprietary systems. However, they do rely on access to the wireless spectrum allocated by the Federal Communications Commission (FCC). The allocated band became more and more valuable and slow progress with V2X deployment – which requires large scale cooperation between the roadway providers and the vehicle suppliers – resulted in a significant reduction in the available spectrum. This cooled enthusiasm all around.

Vehicle manufacturers vary in their enthusiasm for V2X and may have their doubts about road agencies' level of commitment and willingness to invest. V2X is being developed globally, but there are subtle differences in bandwidths and wireless standards. There are also differing approaches to use cases and how messages from other vehicles should be interpreted, assembled and actioned as warnings. One global automaker – the Toyota Motor Corporation – demonstrated unusually strong commitment to V2X and attempted to provide a certain level of V2X safety as an option for customers.

In Japan, Toyota released V2X functionality, including elements of both V2V and V2I, for three models within their range of vehicle offerings. On September 30, 2015 Toyota announced a V2X package called ITS Connect. It included two use cases for collision warnings at intersections and one use case for adaptive, cooperative cruise control between multiple vehicles. This initiative was greatly assisted by Japan's existing endowment of suitable radios in the national road infrastructure.

In addition to several localized model deployments of V2X in the United States, the FHWA launched a V2I precursor initiative called Signal Phase and Timing (SPAT). SPAT fell short of V2I in full, in that it simply provides real-time broadcasting of the phase changes of an intersection's signal controller. However, it is intended that SPAT will be engineered in a manner that will pave the way for a more complete version of V2I at a later time. SPAT was couched as a challenge to all 50 states to fit out at least 20 intersections. According to the National Operations Center of Excellence (NOCoE) (19), about half the states had responded as of May 2024.

4.7.1 THE WICKED PROBLEM OF V2X REALIZATION

Obviously the totally new system capabilities of V2I and V2V bring many challenges in their realization: it is difficult to even indicate the safety benefits with a handful of active intersections and vehicles. But model deployments in the realm of research, such as the U.S. DOT's Ann Arbor Safety Pilot (20), with a few thousand vehicles, have proven the ability of V2X to warn drivers and reduce risk across the board. This research has shown clearly that V2X is the only known road safety paradigm worthy of the label transformational.

Model deployments of V2X have provided new lessons concerning the centrality of privacy and cybersecurity. V2X creates rich data streams that not only allow us to avoid crashes but could potentially throw open people's daily driving lives – perhaps for purposes that are not currently known. The development of V2X began in the early 21st-century era of "data as the new oil" and increasing awareness of privacy and security as intertwined socio-technical issues. The U.S. DOT's architecture for V2X, as it moved through several guises starting with Vehicle Infrastructure Integration (VII), was designed to prevent the identification and tracking of vehicles out on the road. At the same time, vehicles broadcasting their present dynamics with DSRC radios needed some level of authentication.

The solution adopted early on was to issue radios installed in vehicles with rotating certificates, virtual of course, that proved their bona fides without recourse to a permanent identifier. For the purposes of V2X use cases, vehicles also have a temporary identifier that lasts long enough to see it through an intersection, for example, but certainly not from an origin to a destination. Sound cybersecurity measures were also built into the V2X architecture supported by U.S. DOT and underwritten by the auto industry. Along the way, the massive stakes for privacy and security of personal data have rung alarm bells at many levels of society, the media and elected officials. Issues concerning the collection, storage and re-use of personal data have proliferated and large data breaches have occurred.

When it comes to personal mobility data, the main concern still appears to be the ability of the government to access such information and compromise the sense of freedom entailed with driving anywhere, anytime. This type of risk has been well addressed in the design of V2X. Concerns about unauthorized hacking of V2X systems arise through possible compromises of safety, but the lack of financial incentives make V2X a relatively small target for hackers. And AVs are now dominating security discussions in the field of mobility.

As long as humans continue to drive in large numbers, V2X is needed to mitigate collision risks for everyone, everywhere. Reverse engineering from the study of individual traffic incidents cannot seriously reveal and address driving risk throughout the system. We have no systematic way to connect all those dots, each one representing somebody's death or serious injury. For the first time, V2X will reveal our less-safe system and provide the means for a safe system. The wicked problem is to create a single technical capability across the entire national road network. We are faced with a network problem, and leadership is needed on behalf of the whole network. This cannot be achieved using the current network method of franchising. And the automotive and ITS industries can only achieve so much.

4.8   NEW MOBILITY MODES AND ENVIRONMENTS

4.8.1   THE RISE OF E-BIKES AND SCOOTERS

The advent of cheap and reliable electric propulsion grafted onto traditional personal machines such as bikes and scooters has created new categories of urban mobility. With higher speed capability than unpowered equivalents, such machines are more competitive with motor cars in certain environments. The State of California was quick to introduce model legislation that would define several classes of E-bikes, along with several operating environments, plus associated matters such as age requirements, licensing, registration and protective equipment. According to Forbes (21), there are three categories of E-bikes with speed capabilities in the range 20–28 mph. None require licensing or registration and only Type 3 – with the highest speed capability – require the rider to wear a helmet. All have access to several classes of bikeway, with some restrictions for Type 3 bikes.

California rules for E-scooters are simpler. There is a blanket speed limit of 15 mph and a requirement to use bikeways whenever they are available. Riding on sidewalks is banned and riding on roadways is permitted for posted speed limits up to 25 mph. Riders must have a driver's license or learner's permit.

As the popularity of E-bikes and E-scooters has increased, so the rules governing their use have proliferated. So much so that manufacturers have taken it upon themselves to provide summaries and updates concerning requirements across the States. For example, Velotric were offering the following detailed information as of March 28, 2024. E-bikes are not regulated as motor vehicles and come under the Consumer Product Safety Commission (CPSC). The CPSC specifies three tiers, with a maximum power of 1.01 horsepower (750 W) and a maximum speed up to 28 mph provided by the electric motor alone (without pedaling). According to Velotric, about half the states have adopted these three-tier definitions. But other safety requirements vary between the states.

Somewhat in contrast, E-scooters are not subject to CPSC requirements and are generally regulated less formally than are E-bikes. The manufacturer Unagi provided the following overview as of April 24, 2024. There are no federal requirements for power and speed, but most states have a power limit of 1,000 W and specify a maximum speed of 15–20 mph. It is notable that neither registration nor licensing is required, with some exceptions. Several States including California do require a driver's license or learner's permit. Questions abound concerning riding on sidewalks and conditional use of bikeways.

4.8.2 Usage and Potential Harm to Riders

The U.S. DOE (22) reported that annual sales of E-bikes increased by a factor of four between 2019 and 2022, reaching 1.1 million units in 2022. In the case of E-scooters, a high proportion of rides use shared machines. According to the National Association of City Transportation Officials (NACTO) (23), the year 2022 saw almost equal shared trips using E-bikes and E-scooters: 67 million trips by E-bike and 58.5 million trips by E-scooter. These figures are quoted as totals for the United States and Canada combined. Such numbers may be compared with the AAA Foundation for Traffic Safety's 2022 American Driving Survey (24): Americans made 227 billion driving trips of average length 30.1 miles. Therefore total trips by U.S. urban E-machines (shared bikes plus scooters) represented less than 0.06% of trips by car.

Nevertheless, serious injuries incurred riding E-bikes have attracted the attention of the medical profession – perhaps shades of the early days of motor vehicle collision mayhem. Online medical articles such as "ER Visits Due to Electric Bike Injuries Soar Across U.S." posted on April 1, 2024, show familiar concerns about recurrent patterns of serious injury and lack of helmet wearing. The authors quote research based on national electronic reporting of injuries based on emergency room (ER) visits and hospitalizations. In a five-year period (2017–2022) there were 45,000 ER visits and 5,000 hospitalizations related to E-bike crashes and incidents. These numbers suggest that E-bikes, while newsworthy, are doing a small quantum of harm relative to car crashes.

4.8.3 Impact on Street Design

Many cities worldwide are encouraging a transition from motor cars to personal machines including the powered versions – E-bikes and E-scooters. Such encouragement includes rules and guidelines for the machines themselves, rider preparation and behavior and improved street designs that are intended to be more machine-friendly. Organizations such as the National Association of City Transportation Officials (NACTO) have published manuals for street design in the mode of complete streets. They are motivated by streets' occupying more than 80% of public space in cities and their conviction that "they often fail to provide their surrounding communities with a space where people can safely walk, bicycle, drive, take transit, and socialize". NACTO's Urban Street Design Guide (25) provides a compendium of physical elements including speed tables, pinchpoints, gateways, chicanes, bus bulbs and many more. Such designs are in evidence to varying degrees worldwide and represent a significant evolutionary trend in cities and population centers. Given that E-machines currently represent a small contribution to meeting people's overall demands for mobility, their biggest impact may well be in urban designs and layouts.

4.9 STRATEGIZING FOR ROAD SAFETY

4.9.1 INFLUENCE OF VISION ZERO

Subsequent to the release of the global Vision Zero propositions late in the 20th century, a number of countries developed tailored road safety strategies, partly in response to the new philosophy. Ideas that attracted attention relate to road safety as a more complete system with less attention on the vagaries of drivers and more on the design of the road and traffic system. And more attention was directed to the morality of road safety, making it clear that human life could not be exchanged for the economic benefits of driving.

Many were attracted to what may seem like an extremely audacious goal embodied in the title of the initiative: rather than adopting targets to reduce annual fatalities by a certain percentage, could we aspire to having zero? In a forgiving system, drivers' mistakes would not result in their losing their lives. It was made clear that there may not be zero collisions, but there would be zero deaths. In a sense, this harks back to William Haddon's mid-century epidemiology focusing on the "second collision" – that of the driver with the interior of the vehicle – rather than preventing the first collision. It is not surprising that national road safety strategies placed strong emphasis on survivable traffic speeds aligned with specific classes of roadway.

4.9.2 EFFECTS OF THE NEW STRATEGIES

Australia was an early proponent of Vision Zero principles and is an experienced user of national road safety strategies. Currently, total fatalities number approximately 1,000 persons killed. Following several ten-year iterations, the National Road Safety Strategy 2021–2030 (26) injects an increased sense of urgency into the prime task of reducing total national fatalities. The previous decadal version of the strategy called for a 30% reduction (2011–2020), but this had not been achieved; rather, fatalities reduced by a still-impressive 22%. There is a renewed focus on a number of quantitative indicators of safety progress and broader scrutiny of those measures. And the 2030 target is a stiffened 50% reduction in fatalities. The proponents of Australia's strategy are clearly serious about the long-term goal of Zero (fatalities and serious injuries) by 2050.

Australia intends to put its targets into practice via many comprehensive insights and plans. Fundamentally, speed management is a key strategy and is seen as part of safer operation of the infrastructure. Safer vehicles are sought through technologies on an arc from crash avoidance to connected vehicles and to sound preparation for AVs. VRUs are to benefit from lower speed limits, tailored street treatments and greater enforcement. Finally, risky driving – covering the gamut from impairment to distraction to carelessness and evidenced by speeding – is being approached with education and enforcement. The pressure to show results is making sure that Australia's national safety actions remain staunchly utilitarian, with clear targets based on outcomes.

Sweden and Australia were close collaborators on the original development of Vision Zero, and both have followed through with safety strategies, action plans and targets. The Swedish Government decided on the 2030 fatality reduction target of 50%. However, "facing major challenges" (27) they relaunched Vision Zero in 2016 and adopted shorter term Road Safety Action Plans, currently covering the period 2022–2025. The 2030 target remains and equates to 133 persons killed. The current action plan calls up increased speed management, including automatic traffic control (ATK) systems. The plan also targets impaired driving, bicycle safety and pedestrian safety. The current plan makes for gloomy reading and hovering over it is the Swedish Parliament's equally strong commitment to active mobility – principally a strong increase in the vulnerable modes of cycling and walking.

In the United States, the non-profit Vision Zero Network assists many jurisdictions – mainly cities – to implement safety plans based on Vision Zero. New York City launched an extremely comprehensive Vision Zero program in 2014 when there were 259 fatalities on NYC's roadways and streets (28). While there were significant reductions in intervening years, they were not sustained and the number reported in 2022 was virtually unchanged – 260 fatalities.

It is notable that pedestrians were by far the biggest category (numbering 121 fatalities) followed by motor vehicle occupants and operators (61), motorcyclists (38), E-mobility users (21) and cyclists (19). While many laudable programs were launched and completed – well covering the "three Es" (education, engineering and enforcement) – automated speeding summonses absolutely stand out. Since the launch of Zero in New York, there were 20,969,529 such summonses, rapidly rising each year. In 2022 alone, there were almost 6 million tickets issued by speed cameras.

4.9.3 A STRETCH GOAL?

Some jurisdictions have interpreted Vision Zero as requiring road deaths to reduce to zero in the somewhat distant future (for example, by 2050); the Australian approach of ten-year horizons that are re-targeted is a good example. In some countries, progress does not meet expectations; in some cases, the numbers do not change greatly over a period of years. Most jurisdictions then double down on their efforts. Safety programs become much more comprehensive and better managed and generally bring all known effective measures to the table. This works, but not to the extent of realistically approaching Zero.

When scenario analysis is carried out to probe the feasibility of such ambitious aims, it is usually found that massive reductions in impact speeds would be needed, among many other dramatic measures such as new technology penetrating the national vehicle park much faster. Nevertheless, some may regard even limited progress as a safety success – the syndrome of the stretch goal.

In varying degrees, other aspects of the original Vision Zero have influenced safety strategies. The "no trade-off" sanctity of human life may be seen in continued tightening of speed limits, along with much wider surveillance of driving behavior in some jurisdictions. This form of oversight may include the widespread detection of driver actions that are rarely stand-alone causes of crashes; for example, enforcement technology may target non-driving acts that *could* be distracting. But of course no-one knows whether a driver actually is distracted.

4.9.4 SAFE ROAD AND TRAFFIC SYSTEM

The Vision Zero philosophy clearly espoused the Safe System, where a driver's error would not cause them to lose their life. The Netherlands created their own version of the safe system paradigm, called Sustainable Safety. This approach, unlike many other national strategies, broadly entertains avoidance of the first collision as well as moderating the second collision. In its third iteration, for the period 2018–2030 (29), the onus is shifted from the driver to the driving environment. The "full court press" seen in some national strategies is avoided by enunciating key principles for a safe system, covering both the first and second collisions. First, it is recognized that roads do not all have the same functionality, and we should make it much easier for motorists to absorb a road's status and purpose as they go. Second, the road and traffic system should protect humans by following the principles of mechanics, namely, "minimizing differences in speed, direction, mass and size". That is, we should not be neglectful in road design, with everyone pinging around willy-nilly. And third, knowledge of human factors engineering should inform the design of the road traffic environment, to suit human competencies. Perhaps for the first time, road design principles address human protection and do not leave it all to the vehicle.

Over the decade 2007–2017, total Netherlands road fatalities reduced from 791 to 613, a reduction of 23%. It is not clear whether this was the result of fundamental principles – dare we call them golden rules? – applied to road traffic or more organized use of a multitude of known countermeasures. What may be happening is that the golden rules provide more boldness in the application of certain measures that were already in use. For example, vehicle speeds in pedestrian zones are subject to technology for speed limiting. Or personal devices are blocked in order to remove an important source of driver distraction. Policy makers need to feel that they are on very firm ground indeed before they would support such measures. This is where human factors science and engineering comes in, as enunciated in the Dutch literature on Self-Explaining Roads (30). And a safe system built upon such life-enhancing knowledge is going to be more sustainable than one based on restrictions and penalties.

REFERENCES

1.   United States Bureau of Transportation Statistics (2022) Contribution of Transportation to the Economy. https://data.bts.gov/stories/s/Transportation-Economic-Trends-Contribution-of-Tra/pgc3-e7j9/

2.   United States Cybersecurity and Infrastructure Security Agency (2024) Transportation Systems Sector. Highway and Motor Carrier. www.cisa.gov/sites/default/files/publications/nipp-ssp-transportation-systems-2015-508.pdf

3. United States Environment Protection Agency (2023) The 2023 EPA Automotive Trends Report. EPA-420-S-23-002. www.epa.gov/system/files/documents/2023-12/420r23033.pdf

4. New York City (NYC) Department of Transportation (2019) NYC Mobility Report. Aug. 2019. www.nyc.gov/html/dot/downloads/pdf/mobility-report-singlepage-2019.pdf

5. United States Bureau of Labor Statistics (2023) Consumer Expenditures in 2022. Report 1107. Dec. 2023. www.bls.gov/opub/reports/consumer-expenditures/2022/

6. Hardin, Garrett (1968, Dec. 13) The Tragedy of the Commons. *Science, New Series*, Vol. 162, No. 3859, pp. 1243–1248.

7. EPA (2024). Inventory of U.S. Greenhouse Gas Emissions and Sinks: 1990–2022 U.S. Environmental Protection Agency, EPA 430R-24004. www.epa.gov/ghgemissions/inventory-us-greenhouse-gas-emissions-and-sinks-1990-2022

8. Davidow, Bill (2012) The Tragedy of the Internet Commons. *The Atlantic*. May 18, 2012.

9. Swift, Earl (2012) *The Big Roads*. HarperCollins Publishers.

10. Insurance Institute for Highway Safety (IIHS) (2023) *Fatality Facts 2022*. Pedestrians.

11. Governors' Highway Safety Association (GHSA) (2023) Pedestrian Traffic Fatalities by State 2022. www.ghsa.org/sites/default/files/2024-06/2023%20Pedestrian%20Traffic%20Fatalities%20by%20State.pdf

12. Federal Highway Administration (FHWA) (2022) Highway Statistics 2022. www.fhwa.dot.gov/policyinformation/statistics/2022/

13. United States Environment Protection Agency (2023) Sources of Greenhouse Gas Emissions. www.epa.gov/ghgemissions/sources-greenhouse-gas-emissions

14. United States Bureau of Transportation Statistics (2024) Hybrid-Electric, Plug-In Hybrid-Electric and Electric Vehicle Sales. www.bts.gov/content/gasoline-hybrid-and-electric-vehicle-sales

15. Texas Transportation Institute (2019) *2019 Urban Mobility Report*. The Texas A&M Transportation Institute & INRIX.

16. National Highway Traffic Safety Administration (2006) The 100-Car Naturalistic Driving Study. DOT HS 810 593. www.nhtsa.gov/sites/nhtsa.gov/files/100carmain.pdf

17. National Highway Traffic Safety Administration (2007) Pre-Crash Scenario Typology for Crash Avoidance Research. DOT-VNTSC-NHTSA-06-02. www.nhtsa.gov/sites/nhtsa.gov/files/pre-crash_scenario_typology-final_pdf_version_5-2-07.pdf

18. Retting, R A., Ferguson, Susan A., & Hakkert, Shalom (2003) Effects of Red Light Cameras on Violations and Crashes: A Review of the International Literature. *Traffic Injury Prevention*, Vol. 4, No. 1, pp. 17–23.

19. National Operations Center of Excellence (2024) SPaT Challenge Overview. https://transportationops.org/spatchallenge

20. United States Department of Transportation (2024) Connected Vehicle Safety Pilot. www.itskrs.its.dot.gov/sites/default/files/2024-04/executive-briefing/CVPilots_Update_2024_EB_04-18-24.pdf

21. Thompson, Heather & Kennan, Hallie (2017) As Transportation Costs, Emissions Grow, Electric Bikes Offer an Efficient Alternative. *Forbes*, July 6, 2017.

22. United States Department of Energy (2023) FOTW #1321 E-Bike Sales in the United States Exceeded One Million in 2022. Dec. 18, 2023.

23. National Association of City Officials (NACTO) (2023) Shared Micromobility in the U.S. and Canada: 2022. https://nacto.org/wp-content/uploads/2024/08/Shared-micro-in-2023-snapshot_FINAL_July22-2024.pdf

24. Zhang, X. & Steinbach, R. (2024). *American Driving Survey: 2023 (Research Brief)*. Washington, D.C.: AAA Foundation for Traffic Safety.

25. National Association of City Officials (NACTO) (2024) NACTO's Urban Street Design Guide. https://nacto.org/publication/urban-street-design-guide/

26. Australian Infrastructure and Transport Ministers (2021) National Road Safety Strategy 2021–2030. www.roadsafety.gov.au/sites/default/files/documents/National-Road-Safety-Strategy-2021-30.pdf

27. Swedish Transport Administration (2023) Road Safety Action Plan 2022–2025. Document No. 2023-086. https://bransch.trafikverket.se/globalassets/dokument/vision-zero/road-safety-action-plan-2022_2025.pd

28. City of New York (2014) Vision Zero: Building a Safer City. www.nyc.gov/content/visionzero/pages/

29. Dutch Institute for Road Safety Research (SWOV) (2018) *Sustainable Safety 3rd Edition – The Advanced Vision for 2018–2030.* SWOV.

30. Theeuwes, Jan (Ed.) (2017) *Designing Safe Road Systems: A Human Factors Perspective.* Routledge, CRC Press.

5 The Emerging Safety Paradigm of Connected Vehicles and Infrastructure

The new century commenced with some important new engines of road safety already started: the advent of infrastructure intelligence in the form of Intelligent Transportation Systems (ITS) and the bold philosophies and national strategies of Vision Zero. Although ITS was formulated to safely move a lot more traffic on the existing road network, it was ready to become an important source of safety improvement in its own right. And Vision Zero demanded a quantum reduction in fatalities brought about by the conscious creation of a safe system; there would be less emphasis on the responsibility of human drivers.

Road safety in the 20th century had succeeded in addressing all three of the key elements in the transportation system – vehicles, roadways and drivers – in a comprehensive search for countermeasures. The time had come to address the behavior of the system – the way these distinct components come together. A major divide in shared imagination was known to exist between the national cohort of vehicles and the coast-to-coast road network. A new technological age was rendering this lack of system coherence more tangible. It was difficult to make a discontinuous system safer.

A 21st-century cyber-physical system would be created where vehicles could communicate with other nearby vehicles and with the infrastructure they were all traveling on. A specific form of rock-solid wireless communication would be used to continuously convey information in a standard format. Many different digital technologies would be pulled together for yet-to-be-defined system purposes and the result would be more than a technology. The new sensibility was called connected vehicles and infrastructure – CV for short. Hard on its heels, a more compelling technological solution appeared with vehicle automation (AV for short) claiming to be capable of replacing human drivers.

Despite becoming the ugly duckling of these new transportation technologies, CV has been respected all over the world; and it has told us a great deal about the safe system that we do not yet have.

DOI: 10.1201/9781003483861-5

5.1 THE DISTANT RELATIVES OF INFRASTRUCTURE AND VEHICLES

The worlds of motor vehicles and roadway infrastructure are clearly synergistic by intent but have been developed by quite different cohorts, with their own motivations and constituencies. Not the least of many divides is that between the public and private sectors. Those who make and sell vehicles constitute one of the most important economic drivers of the 20th century. The public road network they utilize is provided by government agencies for a multitude of societal purposes, serving the needs of citizens as well as all of industry, supporting services and the military.

Prime among the peoples' roadway needs are transportation of every stripe, housing and many private, public and economic activities in common. Vehicles are designed and operated to meet the capabilities of the road network, and the infrastructure is designed to enhance the mobility of vehicles that conform to established norms. Good design of both means that they traditionally exist at arm's length from each other. One reason for such a distant relationship is that vehicle design moves forward in brisk consumer product cycles, while the design of roadway infrastructure operates on much longer lifecycles.

5.1.1 THE INFRASTRUCTURE REACHES OUT TO VEHICLES

In the early years of the 21st century, with the new field of ITS on the rise – at the same time as industrial wireless communication was coming of age – a new technical movement was born in the United States: it was termed Vehicle Infrastructure Integration (VII).

On February 9 and 10, 2005, lightning struck in San Francisco, when the U.S. DOT invited the tribes of automotive, ITS and infrastructure to come together. The meeting (1) was primed to reveal a fully formed paradigm of system innovation: no less than "an enabling communication infrastructure" to go along with the roadway infrastructure. A nationwide, roadway-based communications network was to be created. At that time, the U.S. DOT already had a vital and growing interest in roadway system management and operations, and VII could become an essential enabler of smarter infrastructure use.

Right from the start, VII was propelled by the two driving forces of safety and mobility. This duality led to a proliferation of VII functions and later created difficulties in portraying clear purpose using widely understood terminology. For this and other reasons, VII and its successors have remained out of the public eye and the technology is not considered a strong commercial opportunity. In 2005, VII was presented as a multi-purpose network, with safety first among equals; government interest in VII was also predicated upon traffic management challenges and traveler information needs. The safety case was put very well by the respected National Highway Traffic Safety Administrator Jeffrey Runge, as follows (1):

> While crashworthiness standards have been and will continue to be very important, we are reaching the point of diminishing returns by focusing only on crashworthiness. The biggest return on investment in terms of lives saved and

injuries prevented in the future will come from accelerated development and deployment of crash avoidance technologies.

5.2 A PROBLEM OF SEMANTICS

5.2.1 THE THICKET OF CONNECTED VEHICLE TERMINOLOGY

Since that seminal year of 2005, VII experienced several name changes, including SafeTrip 21 and IntelliDrive. In later years additional terms entered the professional dialog of what started as VII: these include vehicle-to-vehicle (V2V) communication, vehicle-to-infrastructure (V2I) communication, Vehicle-to-Everything (V2X) communication and Cooperative ITS (C-ITS). Sometimes the same set of activities is called "Connected Vehicles (CV)" or "Connected and Automated Vehicles (CAV)". In the latter case it was often anticipated that the two technologies of connected vehicles (CVs) and AVs would converge. One important reason for retaining separate vehicle and infrastructure terminology was a widely supported push for connected vehicle rulemaking, so that all new vehicles would eventually have to contain the initially chosen standard for V2V (and V2I) radios – Dedicated Short Range Communication (DSRC). Note that the V2V and V2I radios had to be the same, and hence the use of the more general term V2X. But after many years of R&D investment, V2V research and rulemaking ceased in 2017 under a new federal administration averse to expanding vehicle rules.

The term "connected vehicles" (CV) refers to connection that could be used by all ground vehicle players: certainly cars, followed by freight trucks and buses. Eventually motorcycles, bicycles and pedestrians would need to be included to complete the safety paradigm. But CV is not synonymous with V2V because CV also implies connection to the infrastructure. The term "connected" originally referred to licensed wireless regimes epitomized by DSRC; subsequent developments in wireless led to the inclusion of the newer cellular technologies. On the other hand, satellite, internet and forms of telematics would not be included in the meaning of "connected vehicle" because they may lack immediacy and reliability and are not federally licensed. The connected vehicle is designed to support the driver at all times, but the driver still drives the vehicle. Hence the clear dichotomy between the paradigms of "connected vehicle" and "automated vehicle".

In addition to CV technology the driver may also be supported by vehicle-based systems under the auto industry rubric of advanced driver assistance systems (ADAS). Part of the appeal of CV in the early days was its affordability. Because its effectiveness was exponentially related to its density of deployment out on the road, the federal government was attracted to the potential of CV as a low-cost version of ADAS; automakers offered ADAS applications as paid extras. It appeared that the same safety applications – say forward collision warning (FCW) – could be enabled with either ADAS or CV. It turned out that ADAS had a life of its own and could be integrated with existing vehicle controls like brake systems. Automakers were able to develop more complex functionality in ADAS alerts than would be possible with CV applications. Brake signals could be used to suppress forward alerts if the driver was already braking.

5.2.2 THE SPECTACLE OF AUTOMATED VEHICLES

The term "automated vehicles (AVs)" refers to various levels of replacement of human control, applied to a similar range of ground vehicle players as applicable for V2X: cars, freight trucks and buses. Like ADAS, AVs are largely in the province of the auto industry. However, the heart of the AV – the automated driving system (ADS) – would often be a stand-alone product; its manufacturer may not be an auto-maker or established Tier 1 supplier but could be a third-party specialist manufacturer.

Generally, the paradigm of AV would have a much broader impact than CV because it involves adjacent industry sectors such as ITS, telecommunications, insurance and law. While CV has stayed in the background, the development of AVs has leapt into the consciousness of industries and the public alike, with large R&D investment. The high value of AV intellectual property far outweighs that of CV: the realization of CV – making use of existing technologies – is more of an institutional problem while AV has presented a technological mountain for the industry to climb. For different reasons, both paradigms present significant challenges in proportion to the rewarding safety and mobility vistas they open up.

The terms "automated" and "autonomous" are often used interchangeably. However, "autonomous" usually means a vehicle that relies on its own on-board sensors for situation awareness in the roadway, and therefore for exercising vehicle control functions. The term may seem to imply a high degree of automation or self-driving capability but this is not necessarily the case. "Automated" vehicles may cover a very wide range of automated driving features, often falling well short of self-driving capabilities. For example, self-parking would be one example of a low-level automated performance feature. Sometimes it is easier to just say "driverless cars", although this does not help in tightening up the terminology.

5.2.3 CONNECTED, BUT NOT AUTOMATED

When we hear about "connected and automated vehicles", we may have a sense that the two underlying technologies are closely related or even interdependent. In fact they are separate technologies for the most part. The connected part of "connected and automated" generally refers to a V2X-enabled capability that creates machine awareness of the trajectories of equipped vehicles in the immediate vicinity. This machine awareness applies to vehicles as well as specific features of the infrastructure, such as intersections and curves. Machine awareness may be used to identify safety risks, but also to condense or smooth traffic flow. These applications of the technology require warnings and notifications for drivers, in order for the driver to make the required vehicle corrections.

CV only becomes highly effective when the density of equipped vehicles in the traffic stream increases to a certain point. So the rate of deployment of V2X technology in vehicles and infrastructure is a key issue. This requires consideration of "original equipment" fitment in both vehicles and infrastructure, as well as the use of aftermarket technology for existing vehicles and potentially for infrastructure as well.

The question of automatic intervention that falls short of machine driving arises with certain ADAS systems. Lane departure warnings (LDWs) in ADAS may be accompanied by automatic nudging of the vehicle back into the correct lane. This is not driving automation but could be regarded as an advanced means of notifying the driver – the act of conveying the warning – without resorting to audible or visual means. While such actions may tend to blur the line between human and machine driving, CV technology is squarely in the realm of human driving.

In contrast, the "automated" part of "connected and automated" refers to any substantial form of active control on the part of the vehicle rather than the driver. Unlike CV, requiring a cloud of vehicles equipped in exactly the same way, AV is contained in a single vehicle and does not depend upon technological cooperation from surrounding vehicles.

5.3 THE CORE OF THE CV PARADIGM

Connected vehicles (and infrastructure) rely on standardized, ubiquitous wireless communication regimes. The selected wireless regime provides a platform upon which a rich variety of applications may be placed and operated. The requirements of the application – its purpose and scope – largely determine the necessary capabilities of the assigned wireless communication method. In the United States, connected vehicles have been developed primarily for the purpose of deploying safety applications. While safety applications may cover a wide field in terms of specificity for avoiding crashes, the adopted wireless method must accommodate the most demanding applications.

The platform development that was driven by the U.S. Department of Transportation (U.S. DOT) uses DSRC, a form of Wi-Fi that currently uses a 30 MHz slice of licensed spectrum at 5.9 GHz. The signal reliability and short latency of this wireless regime are suitable for highly specific safety applications like FCW, electronic brake light assist and left-turn assist (LTA). But newer versions of cellular technology, such as 5G, may have adequate technical performance similar to DSRC. Religious fervor on the part of DSRC originalists is no longer helpful and other wireless standards are increasingly being tested.

5.3.1 METHODOLOGY OF V2V

The DSRC (or other) wireless system is used to send Basic Safety Messages (BSMs) that contain vehicle position, speed and heading, as a minimum. BSMs are sent ten times per second and are received by equipped vehicles and infrastructure equipment in the immediate vicinity. This is the single defining immutable mission of V2V. Each host vehicle assembles the streams of BSMs that they receive to create a continuously updated map of vehicles nearby within the traffic stream. Such automated situation awareness would allow the driver of the host vehicle to conduct themselves in an accommodating manner and to avoid emerging tendencies for contact with other vehicles. Such tendencies may develop into escalating risk of collision.

The technologies hosted by the V2V-equipped vehicle include the DSRC system, application software and driver interfaces. Connections to the vehicle's controller area network (CAN) bus are preferred and greatly widen the range of information that may be transmitted – far beyond the BSM – and therefore, the scope and effectiveness of consequent V2V applications. The computational requirements in the vehicle are modest. Each host vehicle needs to be fitted with a Global Positioning System (GPS), plus effective antennas for both the DSRC wireless and the GPS. Continuing availability and reliability of non-profit spectrum in the 5.9 GHz public safety band is an important issue, and the Federal Communications Commission (FCC) has auctioned off part of the initially allocated 75 MHz band, reducing ITS applications to a lesser band of 30 MHz.

As its name suggests, DSRC has a short range – typically a few hundred yards – and the standardized transmission packets that are sent ten times per second may be inadvertently blocked, absorbed or reflected by large trucks, buildings or roadside objects. Problematic locations include so-called urban canyons where GPS signals may drop out. Heavy tree cover in the roadsides also causes signals to drop out. Reliability of transmission may also depend on the number of vehicles transmitting in the immediate vicinity.

The U.S. DOT was forced to cancel rulemaking for the provision of V2V at the original equipment vehicle stage. That rulemaking applied only to the presence and capability of the DSRC communication system, but not to specific applications, the nature of which was left to automakers. As a first step the U.S. DOT wanted to ensure that the transmission packets sent by a vehicle of certain make will be seamlessly received by a vehicle of another make. By requiring CV radios of a certain standard in all vehicles at point of manufacture, it was intended that road agencies would reciprocate by installing compatible roadside equipment. Much momentum was lost when the leading edge of CV rulemaking collapsed.

5.3.2 SAFETY PARADIGM OF V2X

In order to become operational, V2X relies on several technologies, including wireless and GPS. But V2X itself is a *technological safety paradigm* in that it is a model for collaboration, deployment and rhetoric. Once the technological capability was named and widely disseminated, interested parties from government, industry and academia envisaged beneficial use cases that came together in a compelling narrative. V2X's main stock in trade is safety interventions. Sometimes it feels like a movement that needs promotion. The many talents of V2X have been workshopped with knowledgeable professionals around the world. But we may struggle to describe and explain it; this is an important task in itself. The paradigm of V2X cannot exist without collaboration and deployment on a sufficiently large scale, and such constructive new-world building will not even get started without a rhetorical modeling of the V2X experience. Of course the latter problem is also evidenced by the lack of linguistic clarity in the enterprise of V2X. This is also true for the entire field of technical safety. We do not wish to limit ourselves and tend to keep an open mind.

In our time, Erik Hollnagel (2) has questioned the scientific field of human safety – as centered on Hollnagel's interests in industrial and mining safety – and would prefer not to call it safety. His idea of "Safety-II" abandons the definition of safety as "avoiding that something goes wrong" and changes it to "ensuring that everything goes right". It may pay us to step back and consider the paradigm of road safety technology through a lens of qualitative research (3), as distinct from the quantitative research methodologies with which we are much more familiar. Road safety technology seeks to inject more reality into the driver's situation awareness. There are multiple ways that humans interpret reality. The constructivism of Jean Piaget holds that knowledge is developmental and non-objective. He said that "humans cannot be given information which they immediately understand and use; instead humans must construct their own knowledge" (4). This would argue for a deeper meaning for V2X beyond the triggering of warnings.

As drivers we experience our physical world at its most dynamic, abrupt and unpredictable. But we tend to treat it like the floating observers at GM's Futurama. The reality on the ground, in the lane, at the intersection far exceeds that which we could aspire to know. So what subset of that reality do we need to know in order to proceed safely? Even for experienced drivers, it would be miraculous if we observed, deduced and acted upon the right things, all of the time. And everyone around us wears very different sets of "driving goggles". And there are far too many of them entering our orbit and also exiting. So beyond the alluring methodologies of V2X is the unspoken belief that new questions of situation awareness need to be asked and new information transmitted. That information needs to be immediately understood by drivers. Our belief is therefore that knowledge of situation awareness should be a lot more than isolated warnings of imminent contact between vehicles. Because the human is incapable of all the sensing and processing required, technology is required to step in. The driver should be treated according to human factors principles and provided with information pertinent to their front-line responsibilities: anticipating the demands of the roadway system, deciding upon accommodations for other road users and skilled control of the vehicle.

5.4 DEPLOYMENT OF V2X ACROSS THE UNITED STATES

5.4.1 National Roadway Safety Strategy 2022

On January 27, 2022, Transportation Secretary Buttigieg launched the U.S. version of Vision Zero – the National Roadway Safety Strategy (NRSS) (5). Rather than being likened to an epidemic – as in the 1960s – road safety was described as a crisis, with national fatalities experiencing an uptick over several years. The strategy crosses over the modal administrations, including Federal Highway Administration (FHWA) and NHTSA, and adopts a safe system structure aimed at zero fatalities, similar to that adopted in other countries: the strategy contains pillars of safer people, safer roads, safer vehicles, safer speeds and post-crash care.

Commitment is also made to safe system principles, including a focus on reducing fatalities and injuries rather than crashes. And greater mention is made of the responsibilities of road designers alongside drivers and vehicle manufacturers. Attention is drawn to the classes and types of roadways that host fatalities involving motorists and pedestrians, with arterial roads topping the list. In 2020, arterials saw 21,295 fatalities, compared to 12,259 on collectors and local streets, and 5,129 on interstates.

More than half of all fatalities (55%) currently occur on arterial roads. Arterial roads attract a lot of traffic that travels at reasonably high speeds with regular intersections and pedestrians wanting to cross.

The source for infrastructure safety initiatives under the NRSS is a regularly updated and more innovative Manual on Uniform Traffic Control Devices (MUTCD). This manual is produced and supported by FHWA's National Traffic Control Devices Program. It is a standard rather than a set of legislated rules. Adherence to the manual may differ across the 50 states; some simply adopt the manual as is, while some adopt it with certain addenda and some produce their own state version of the MUTCD.

The NRSS continues a significant reliance on vehicle safety technology. Even though NHTSA's rate of rulemaking has slowed over the years, the NRSS calls for at least two important new rules: Automatic Emergency Braking (AEB) and Pedestrian Automatic Emergency Braking (PAEB).

5.4.2 NATIONAL V2X DEPLOYMENT PLAN

Subsequent to the launch of the NRSS, a national vision for V2X deployment was released by U.S. DOT (6). In full recognition of the scale and cooperative nature of the task, the U.S. DOT has stepped forward to take a coordinating, facilitating and supporting role. This role seeks to bring together the efforts of the federal government, the public sector and private industry. The plan covers three relatively short horizons out to 2034. It deploys technology via the National Highway System, 75 major metro areas and the nation's signalized intersections. It envisages six Original Equipment Manufacturers (OEMs) offering 20 equipped vehicle models, selecting from 20 certified V2V radio sets. Safety functionality is seen utilizing five use cases served by the 5.9 GHz public safety band across all 50 states, plus a further five use cases operating outside that band available in five states.

The vast scope implied by the V2X paradigm is indicated by full coverage of the National Highway System plus a high percentage of signalized intersections throughout the country, and even higher in major metropolitan areas. And the deployment plan covers all 50 states. The plan appears to have the infrastructure and vehicle sectors joining hands from the start but is designed to provide majority coverage of the national infrastructure before there is proliferation among vehicle makes and models. And the number of universally available use cases is very modest.

Clearly, the V2X safety paradigm has survived many changes in ownership and supporting technical environment over the years since VII was first proposed. It appears that the initiative has swung back to the infrastructure side of the house, with FHWA taking actions to bring V2X to the forefront for highway agencies across the 50 states. For example, the U.S. DOT has adopted a short-term goal of V2X deployment in 20% of the National Highway System and within the intersections of the top 75 metropolitan areas. But the confirmed role of the U.S. federal government is currently no more than developer and keeper of a desired national roadmap for V2X coverage.

5.5 V2X USE CASES INTENDED FOR COLLISION AVOIDANCE

NHTSA's long-term commitment to crash avoidance, as the next step after occupant protection, had a prominent place in the formulation of VII, as launched in 2005. Use cases for intersection safety included Signal Violation Warning and Stop Sign Violation Warning. These remain quintessential V2I applications that (1) cover a large element of death and serious injury and (2) cannot be done with vehicle-based safety systems.

If all signalized intersections were fitted with standard V2X radios, opting for the same radio in a new vehicle purchase would make great sense to consumers – providing a tremendous reduction in personal risk of injury or death. But NHTSA's limited jurisdiction – vehicle safety standards – placed the boot on the other foot. In order to get the shared V2X paradigm rolling, one side needed to make the first move. VII was always an infrastructure-based initiative, and its stepchild V2X also needs to grow outward from the infrastructure side. But the necessary initiative from the infrastructure side was resisted for a long period while it appeared that the vehicle industry was going to be coerced to make the first move.

5.5.1 THE ANN ARBOR V2X FIELD OPERATIONAL TEST (FOT)

The U.S. DOT foreshadowed a field operational test (FOT) when VII was launched in 2005, and this eventuated in 2012–2013 when the Safety Pilot Model Deployment was carried out in Ann Arbor, Michigan. The test conductor was the University of Michigan Transportation Research Institute (UMTRI). The 12-month FOT was launched on August 21, 2012. The four safety use cases tested in the FOT were (7) as follows:

- FCW; according to NHTSA, this use case "warns drivers of stopped, slowing or slower vehicles ahead";
- Intersection Movement Assist (IMA), which "warns drivers of oncoming cross traffic at an intersection";
- LTA, which "warns drivers of the presence of oncoming, opposite direction traffic, when attempting a left turn"; and
- Blind-spot and lane-change warning (BSW/LCW), which "alerts drivers to the presence of vehicles approaching in their blind spot".

Note that these safety cases differ significantly from those originally proposed for VII (see above) and reflect the differing perspectives of two communities of expertise: the infrastructure professionals who influenced the initial VII model and the vehicle professionals who were the main drivers of the FOT.

As an important part of its collision avoidance program, NHTSA and UMTRI had carried out an intervening FOT called Integrated Vehicle-Based Safety Systems (IVBSS) that tested some of the same use cases but used different technical means to achieve similar functionality. The four safety use cases tested in the IVBSS FOT were (8) as follows:

- FCW;
- Curve-speed warning (CSW);
- Lane-change/merge warning (LCM); and
- LDW; this covered vehicles drifting into unoccupied areas, as well as vehicles drifting toward other vehicles or objects.

Generally speaking, the vehicle-based system provided more accurate conflict detection than V2X. That is why ADASs have endured and prospered. The big exception for the vehicle-based use cases was FCW because activity in the adjacent traffic lane tended to be picked up and misinterpreted. This tended to happen at forward ranges around 50 meters. The lateral warnings LDW and LCM performed extremely well in all respects. They were more valid according to the researchers' "ground truth" systems; they were also responded to faster, and with more active driver corrections. Drivers' absorption of warnings to reduce personal risk was evidenced by much stronger lane corrections when their vehicle was drifting left – toward oncoming traffic, rather than away from it.

The staging of the Safety Pilot FOT benefited from the accurate, reliable vehicle-based data gathered in the IVBSS FOT. It was no surprise that the FCW case using V2V had problems similar to those found with the ADAS version. For one thing, it was necessary to suppress warnings that arose after the driver pressed the brake pedal because they were probably already responding to the problem. Still, only 33% of FCW warnings were issued for valid in-path targets. Most of the false alerts were caused by vehicles one lane over, or in the adjacent left-turn lane. On the flip side, there were zero missed FCW alerts through the course of the year's FOT driving; that is, in no case was there a slower target vehicle in the FOT vehicle's lane without generating an alert. FCW does not fail to warn but it does tend to over-warn drivers.

The V2X FOT included important use cases that could not be formulated using vehicle-based systems alone. The IMA use case posits an unknowing "host vehicle" that could be struck by a "target vehicle" traveling on the cross path and not visible to the host vehicle. Key parameters of this application are the speed of the target vehicle and its distance upstream from the intersection at the moment when the driver of the host vehicle releases their brake and begins to move forward. In order to reduce nuisance alerts, a cut-off was introduced with a maximum time to intersection of the target vehicle of around four seconds. Importantly, there were zero missed alerts for the IMA use case. That is, the driver of the host vehicle was always alerted when a crossing target vehicle was about to appear. But roadway intersections involve many complications. The Achilles Heel of the IMA use case proved to be overpasses which could trigger unwarranted alerts. And a number of further situations can have a target vehicle momentarily on a collision course with a host vehicle, but the real-world scenario is hard-wired so that collision course is not going to materialize. In addition to overpasses, the FOT detected such fanciful sensing for opposing directions (in intersections and curves), cloverleafs, adjacent roads and traffic circles.

The performance of the blind-spot and lane-change use case was encouraging in that more than half of the alerts detected target vehicles approaching in the adjacent lane. Unintended detections included vehicles in the same lane as the host vehicle, or two lanes over. Many of the in-lane false alerts occurred when the host vehicle operated its turn signals while approaching an intersection. This use case experienced one missed alert when the host vehicle operated the turn signal to indicate a lane change and the target vehicle was immediately adjacent – in the immediate zone of risk. In this one case, the host vehicle should have been warned but was not.

5.5.2 CONVERTING CRASH TYPES TO CRASH AVOIDANCE USE CASES

Road safety professionals are strongly aligned to the mentality of countermeasures. What kind of crash are we trying to prevent, and how is it classified? The required crash definitions are limited by the available post-crash evidence and how that evidence is investigated. Designed to marshal the facts on crash causation, NHTSA's comprehensive 2007 study of pre-crash scenarios (9) was specifically intended to inspire and assist the development of crash avoidance technologies. Released around the time of the very early stages of V2X thinking, the pre-crash scenarios are oriented to vehicle-based alerts rather than V2X use cases.

Rear-end collisions are an important crash type, with many serious injuries and fatalities as well as lesser severities. The vehicle-based "forward collision warning" and the same use case mechanized using V2X may seem like a straightforward occurrence at first. But NHTSA came up with three pre-crash scenarios that could fit this use case: (1) lead vehicle stopped, (2) lead vehicle decelerating and (3) lead vehicle moving at lower constant speed. And the former two scenarios were among NHTSA's top five, out of 37, in terms of frequency and severity of outcome. Note that NHTSA articulates each scenario and adds causal factors that had above-average presence in that scenario.

Scenario (1) is described by NHTSA as: "Vehicle is going straight in an urban area, in daylight, under clear weather conditions, at an intersection-related location with a speed limit of 35 mph; and closes in on a stopped lead vehicle." Over-represented factors are given as rural area, driver inattention and exceeding the speed limit.

Scenario (2) is described by NHTSA as: "Vehicle is going straight in an urban area, in daylight, under clear weather conditions, at a non-junction with a speed limit of 55 mph or more; and closes in on a lead vehicle moving at lower constant speed." Over-represented factors are given as driver inattention and exceeding the speed limit.

Use cases intended to protect against these scenarios – whether vehicle-based (ADAS) or V2X-based – encounter great difficulties with false alerts. There is a technical problem in sensing the difference between a slower vehicle in my lane and one in the next lane over. The V2X FOT found this to be a bigger problem the further in front the leading vehicle was, especially around 50 m out.

V2X is, of course, a major nationwide paradigm with many dependencies. On the other hand ADAS systems are a consumer item subject to continuous and rapid development. Because they are factory-fitted, they also have the advantage of potential integration with vehicle controls such as brakes and steering, plus continuous availability of vehicle motion data such as deceleration. This dynamic integration with what the driver is doing and how the vehicle is responding means that a V2X-based FCW use case is going to struggle to compete with the ADAS-based version. If the use case triggers when the host vehicle is very close to hitting the target vehicle, the chances of unwarranted sensing of vehicles in adjacent lanes reduce, but more automatic intervention is needed to reduce reaction times and provide sufficient deceleration. ADAS systems are good at this.

The V2X-based use case known as IMA would potentially counter three of NHTSA's pre-crash scenarios:

- Straight crossing paths at non-signalized intersections;
- Running red light; and
- Running stop sign.

These scenarios all tend to involve lower-speed zones. With respect to roadway environments, the red light running scenario occurs in urban locations, and the others in rural locations. It should be noted that NHTSA does not make any specific connection between the pre-crash scenarios and the IMA use case. However, anticipated crash location may be an important consideration in the deployment of V2X progressively through the network. The Ann Arbor FOT found IMA to be an effective use case in that it warns when it should.

The V2X-based use case for protected lane changes, merges and departures experienced problems with phantom detections in adjacent lanes. The vehicle-based version of this use case had some similar problems but performed better overall and was well-respected by drivers. The NHTSA pre-crash scenario corresponding to this use case is called "vehicle(s) changing lanes – vehicles traveling in the same direction". NHTSA describes the scenario thus: "Vehicle is changing lanes in an urban area, in daylight, under clear weather conditions, at a non-junction with a posted speed limit of 55 mph or more; and then encroaches into another vehicle traveling in the same direction". The NHTSA report adds that driver inattention also contributes. NHTSA had intended for this scenario to be dealt with using vehicle-based ADAS, and currently tested V2X use cases provide little or no advantage.

The pre-crash scenarios developed by NHTSA are still sketchy and do not allow narratives to be constructed or realistic visualizations to be made of key crash types. In terms of causing severe harm, rear-end crashes are one of the most important categories, and NHTSA's intention for automated emergency braking rulemaking will make a large safety contribution, avoiding technical issues with the FCW use case implemented in V2X. However, the V2X use case of IMA has the potential for harm reduction of a similar magnitude but needs to be well-tuned; the logic of the use case is extremely important and needs to be of high quality. As tested in Ann Arbor it is incomplete: it needs to specifically encompass the important NHTSA scenario of vehicles turning at non-signalized intersections. The expanded and tuned IMA use case could well, all by itself, be sufficient to justify the U.S. deployment of V2X.

5.5.3 Multiplicity of Use Cases

Knowledge of human interaction with V2X use cases is not well-advanced, but the opinions of the Ann Arbor FOT participants provide some guidance. Even though the Ann Arbor V2X experiment was set up to generate two levels of warnings – an awareness level and a crash-imminent level – it was found that FOT participants generally dismissed the former as a nuisance. It became clear that drivers dislike being distracted by events they've already noticed themselves and have decided to disregard, and they are definitely allergic to false alerts. Drivers' aversion to false alerts was already known from previous FOTs testing vehicle-based warnings.

NHTSA's progression with crash avoidance use cases started with individual vehicle-based applications and then moved to grouped applications. Individual applications have the advantage that the driver knows exactly what the alert signifies. When the vehicle is fitted with multiple use cases, it may be that two or more applications are triggered at the same time; in the IVBSS FOT, a hierarchy was introduced so that only one warning would be delivered at a time. However, the method of integration of warnings – the I in IVBSS – may be specific to the grouping of use cases.

When NHTSA moved from vehicle-based systems to V2X, the potential number and range of use cases increased significantly. However, at least some leading safety use cases generate alerts so infrequently that – when the big day arrives – the recipient may be startled or confused. The V2X FOT participants varied greatly in their responses to alerts and there was some grouching about the short duration of the actual warning, compounded by its rarity, and the need to pay attention to a mid-mounted screen on the dashboard, read the message and figure out what it had meant.

5.5.4 U.S. DOT Research Findings from the Ann Arbor V2X Model Deployment

The FOT was conducted by the UMTRI. It was UMTRI's job to collect quality data on a sufficient scale, and then turn it over for independent analysis (10) by the U.S. DOT's affiliated experts, the Volpe Center in Cambridge MA. Volpe analyzed 2,384 alerts, only 13 of which caused negative or unnecessary actions by drivers. And none of these reactions was considered to be unsafe.

A major part of the FOT was directed toward understanding driver acceptance of V2X warnings. This was treated as a complex and potentially obscure matter. Acceptance could be reflected in a wide range of driver opinions, understandings and attitudes. For the purposes of the Safety Pilot FOT, the following factors were analyzed to determine driver acceptance: usability, perceived safety benefits, understandability, desirability, security and privacy.

UMTRI researchers had probed the experiences and opinions of a representative number of FOT participants. The FOT was interested in driver acceptance of the entire paradigm of V2X alerts; this would include the intended use case, circumstances under which alerts occurred and the alert process, including the person–machine interface. There were 127 participants who drove 64 fully equipped V2V vehicles, plus a number of partially equipped vehicles; there were also several thousand vehicles which were equipped to broadcast BSMs but were not able to receive alerts. Of the 127 full participants, there were some variations in their exposure to specific use cases. In the case of the IMA use case, 37 representatives were surveyed for their acceptance of that particular mode of alerts.

Following statistical analysis of representative responses, the Volpe Center concluded that many participants approved of the research effort and the well-intended technology. In general, representatives appreciated being alerted to situations of which they were not aware. They also felt safe knowing the vehicle would warn them. Some felt that warnings placed them in a higher level of alertness. On the flip side, many disliked false alarms; some may also experience warnings as distracting or startling and may be confused by certain warnings when they are activated infrequently. It is clear that infrequent warnings work against intuitive use of the system.

In heavy traffic, it may be hard to determine which vehicle was the "target" vehicle, or what the warning was for. Warnings were not appreciated at moments of high driver workload – for example, merging onto a highway. In cases when warnings seemed weird or wrong, participants could waste time trying to figure out what had just happened. Warnings were sometimes distracting, but equally they may be experienced as being startling. For some, a form of anxiety could arise on busy streets, waiting for an alarm to go off. Despite prior expectations of the research team, there was little evidence that users developed an unhealthy dependence upon the V2X technology. Rather than suffering from over-reliance on the warnings, representatives reported having heightened awareness and felt more conscious of safety.

In this FOT, crash-imminent alerts went off at an average rate of 3.2 per thousand miles. As with most forms of naturalistic driving studies, averages have only limited meaning because individual differences between drivers are extremely large. This was also true in the V2X FOT: Volpe found that the most alert-prone driver experienced warnings an order of magnitude more often than the drivers who were least prone. But the alert rate also depends greatly on two factors: (1) how the use case is set up – the degree of proximity to an actual crash before the alert is triggered – as well as (2) the frequency with which equipped vehicles encounter each other.

In the Ann Arbor test, equipped vehicles represented a very small percentage of a typical traffic stream. In any case, crash proximity is a rare occurrence. As the warning threshold is dialed up, and less-critical encounters are suppressed, nuisance levels diminish. However, the system then runs the risk of receding in the driver's consciousness and suddenly appearing in a distracting way, rather than in a constructive way.

The philosophy of the V2X paradigm is just as important as its technical abilities. Many in the safety fraternity may expect such a paradigm, which is technically "always on", to also create a continuous safety mind-set in most drivers. There was good evidence of this in the majority of responses of V2X FOT users, for example:

- "It put me in high alert and made me more observant";
- "I felt utterly safe in this vehicle. I liked the sense of knowing it would warn me";
- "It became a normal part of driving";

- "Makes you keenly aware of vehicle safety";
- "For me, I became more attentive to the cars around me, more observant"; and
- "A driver would pay more attention to avoid warnings".

Much of the emphasis with the V2X FOT centered on the specifics of use cases in avoiding particular crash types. But there seems to have been a significant awareness among participants of an unexpected aspect of V2X – as an arbiter of safe driving. V2X was mainly created to help drivers avoid key crash types, not necessarily to highlight instances of unsafe driving. And yet the technology seems to increase drivers' consciousness of their weaknesses and drivers may even regard alerts as driving failures.

5.6   DEPLOYMENT OF V2X THROUGHOUT THE NATIONAL ROAD NETWORK

The V2X paradigm needs to be capable of rapid scaling across the country because the nation's drivers need to have a consistent appreciation of V2X interventions; it is essential that drivers' expectations are satisfied to a high level. The deployment of V2X 5.9 GHz DSRC roadside equipment in the infrastructure – especially at intersections – is a necessary development, along with a sufficient density of equipped vehicles in a given traffic stream. The roadside equipment is typically housed in traffic control cabinets, with transmission equipment high-mounted on signal arms, poles and gantries.

While equipment vendors and the traffic control industry have the necessary technology, the business case for paying for the installation, operation and maintenance of the roadside equipment is not at all clear. Even the more limited paradigm of V2V communication requires a minimum level of roadside equipment for connected vehicle support functions – such as the operation of security and privacy systems.

Several model deployments have been developed around the country, starting with Ann Arbor (10). The federal government then carried out additional projects in diverse geographic, demographic and economic locations: in New York City, Tampa and the state of Wyoming. These efforts showed that the V2X paradigm could find expression in many different environments and would attract developers of local use cases. Such model deployments usually involve partnerships between state agencies, cities, consultants, universities and companies.

In addition to the roadside equipment, it is necessary to have a "data backhaul" (getting data to a point from which it can be distributed over a network) to centralized locations. This data appears to have great potential from the perspective of road managers. The data, and the ability to communicate securely with vehicles, also offers certain untapped commercial opportunities. Considered on a per-vehicle basis, the costs for vehicle and infrastructure deployment are modest. And requirements for standards development, testing and compliance are modest.

The successful deployment of connected vehicles on a nationwide scale will require a highly strategic and coordinated public–private effort, along with strong decision-making by a range of companies from the automotive, traffic control, infrastructure and the technology industry. Deployment of connected vehicles will be a more demanding institutional, rather than technological, endeavor. And the potential constituency for V2X extends beyond road safety: connected vehicles and infrastructure also offer major improvements in traffic flow, intersection efficiency, energy wasted in stop–start operations and vehicle emissions.

5.6.1 TECHNOLOGIES AND ALGORITHMS FOR V2X USE CASES

Specialized hardware, firmware and software are required to make V2X function. V2I and V2V systems comprise on-board units (OBUs), roadside units (RSUs), V2X applications that process messages and devise driver warnings, various interfaces and data backhaul. Considerable R&D effort has been devoted to V2I and V2V technology and applications. Work led by the FHWA and the Crash Avoidance Metrics Partners (CAMP) has developed a range of V2I applications. CAMP had also pioneered the V2V use cases tested in the Ann Arbor V2X FOT.

Work by FHWA has developed the connections between RSUs and traffic controllers; ITS and controller companies have developed Signal Phase and Timing (SPAT) products. The formats of both the SPAT Message and the MAP Message have been standardized. These multi-sourced pieces of functionality are compiled by specialist companies who supply V2X radios to both road agencies, on the one hand, and automakers and Tier 1 suppliers on the other. Altogether, the government and automakers spent over $1 billion over a decade to advance DSRC and V2X technologies.

V2X alerts are produced by a complex cyber-physical system that combines sensors, computation, data storage, radios, data from the vehicle CAN bus, communication links with external systems and man–machine interfaces. With respect to V2I, algorithms have access to signal controller and traffic sensor data; for V2V, algorithms can reach into the vehicle's CAN bus and reach out for GPS signals.

The U.S. DOT has often taken a lead role in developing and improving V2X algorithms that turn a certain set of situational inputs into alerts in a form useful to drivers. A V2X algorithm envisions a certain scenario, or set of scenarios, that entail emerging risk of collision. The scenarios are selected to map to known crash types, usually of high frequency and high severity. When the scenario has reached a pre-set level of proximity to crash, an alert is generated by the algorithm. In 2007 – at the dawn of V2X – NHTSA had published a compendium of 37 pre-crash scenarios; this compendium was an open invitation to developers of crash avoidance applications.

The NHTSA compendium (9) does not stray too far from its origins in the General Estimates System (GES) crash database. This database contains all 6,170,000 police-reported crashes for the year 2004; it is not a sample, but a complete census. It is clearly an ambitious undertaking to reduce such a large number of crash types to a manageable number of unwanted narratives. The developers of V2X algorithms have gaps to fill in when they imagine the scenario they are trying to prevent. The task is therefore demanding, requiring specific knowledge and skills. The use cases that V2X algorithms address are not mandated and their methodologies are not prescribed, beyond certain standard technical elements such as DSRC, BSM, SPAT and MAP.

From the early days of V2V use cases, NHTSA has relied upon the expertise of CAMP, located in Farmington Hills, MI. CAMP is a floating consortium of automakers. It was originally founded in 1995 by the Ford Motor Company and General Motors to carry out pre-competitive research in active vehicle safety. CAMP has maintained a long association with NHTSA and has also been entering into a series of cooperative agreements with FHWA in the V2X era of the 21st century.

In 2017, CAMP published a detailed account of the V2I work it had carried out for FHWA (11). The V2I safety use cases that received the most attention were as follows:

- Red Light Violation Warning – given my current speed, do I have enough time to stop at the upcoming red light?
- CSW – am I traveling too fast on this curve, given visibility, weather conditions and road geometry? and
- Reduced Speed Zone Warning – slowing down to allow for roadworks or lane closures.

CAMP developed objective test procedures for each of these applications, including allowances for varying conditions. The Red Light use case is particularly challenging because it has to interface with different types of signal controllers; it requires solid information on the map of the intersection and the logic and timing of phases.

The Red Light V2I use case is based on straightforward good practice: when the light turns red, do I have time to stop? This use case operates from the point of view of the driver but its purpose is conceived from the perspective of the roadway designer. It is within the purview of the infrastructure operator to have vehicles stopping at intersections in a predictable manner, and making way for cross-traffic. But this rule of thumb for safe driving does not map directly with NHTSA's pre-crash scenarios that involve running red lights or rear-end collisions. Of the 37 pre-crash scenarios presented by NHTSA, there are two specific to signalized intersections:

- Running Red Light
  Vehicle is going straight in an urban area, in daylight, under clear weather conditions, with a posted speed limit of 35 mph; vehicle then runs a red light, crossing an intersection and colliding with another vehicle crossing the intersection from a lateral direction.
- Following Vehicle Approaching a Stopped Lead Vehicle – "Vehicle is going straight in an urban area, in daylight, under clear weather conditions, at an intersection-related location with a posted speed limit of 35 mph; and closes in on a stopped lead vehicle". "This typically happens in the presence of a traffic control device".

Those who run red lights therefore run a serious risk of experiencing a right-angle collision with another vehicle, but this is not covered by the CAMP V2I Red Light alert. And the largest group of high-severity crashes at intersections are rear-enders, but they are also excluded from cover by the V2I Red Light alert. The Red Light alert therefore could not truly be described as a crash countermeasure: it is better described as a risky driving alert. It is an extension of the traffic engineer's desire for the vehicle to stop at the intersection, as typically expressed in the red signal itself.

There is a dilemma in helping a driver to avoid a particular type of crash, as against ensuring that the driver is paying attention when a red light is imminent. As it happens, serious rear-enders result when the second vehicle fails to stop. Ironically, the "good practice" version of the alert – aimed at the first vehicle – also serves to avoid a frontal crash for the second vehicle. Could it be that humans driving toward an intersection in a state of alertness are as good as it gets for crash avoidance?

If we pay attention to the opinions of participants in the Ann Arbor V2X FOT, they are not always interested in the details of individual use cases but are amenable to reminders of good practice. A driver in a state of uncertainty about when the current green phase will turn red is not thinking about any particular crash sequence. They currently don't know how soon the red will appear. They may also entertain the option of continuing through the intersection after the lights have changed. Once they get the alert, everything changes. They are now confident about their need to stop and also know that they can stop in time. Importantly, they can discard any notion of continuing through the red signal; in doing so, they avoid the high risk of a right-angle collision. As a bonus, if the vehicle traveling in the lane behind them is V2I-equipped, they will not be rear-ended when they do stop. This example illustrates how the paradigm of V2X – in this case becoming operational via V2I technology – can counter inattention and indecision, and sever causal paths to collisions at signalized intersections.

There is an important distinction here. V2I and V2V can be configured as a generator of crash countermeasures whereby alerts are provided to drivers. In that case, drivers need to have an immediate appreciation of the reason for the alert and, if appropriate, which nearby vehicle they need to avoid. Alternatively, V2X can serve as a human factors helpmate. In the case of CAMP's Red Light alert, V2I provides more information on the intersection's demand for them to stop: that they need to do so immediately and if they do so they will have time to stop in a stable manner. The intersection now knows the location and speed of the vehicle, as well as the layout and signal phasing of the intersection. The all-important connection removes the doubt in the mind of the driver and allows them to make their decision to stop rather than taking a big risk to proceed on the red.

5.7   THE EXPANDING FIELD OF TECHNOLOGICAL SAFETY INTERVENTION

5.7.1   Assistance, Alerts, Warnings and Interventions

As 21st-century safety technology has developed and has been applied in vehicles and in the infrastructure, it has acquired considerable momentum. The sheer velocity of devices, interactions, combinations and layers has increased the rate of innovation. Such technologies start out relevant and work to become pertinent. Some may become compelling. They are a means to an end – helping drivers avoid crashes. ADAS is an important part of our rising capabilities in pre-crash intervention. And some ADAS capabilities are subsequently mandated in new vehicles. Meanwhile, V2X brings similar capabilities, but not exactly the same. And the connectivity of personal devices carried into vehicles, and their traffic applications are also coming closer to the provision of safety advice.

However, knowledge of crash causation with human drivers in the loop has been much slower to advance than the technology. It is therefore difficult to be precise about definitions, objectives and assessments. Nothing tends to be ruled out simply because it is not firmly positioned on the spectrum of practicality. Technological intervention is here to stay, but what exactly is the purpose? And where does it operate along the timeline from a risky situation, to a driver alert, to warning of an imminent crash?

NHTSA's 2007 compendium of 37 pre-crash scenarios (9) covers 99.4% of all light vehicle crashes in the year 2004. Its quantification of human years lost, frequency of occurrence and cost is extremely comprehensive. Its description of pre-crash scenarios goes from crash type (what was the vehicle movement at the time?) to scenario (what was the driver doing relative to the environment?) to causal influences (what helped turn risk into adverse reality?). While all of these pieces of information are essential for technological countermeasures, the GES census provides only bare-bones insights. In a nutshell, the scenarios could be described as a worthy attempt at "situation awareness gone wrong". Among other objectives for this major research effort, NHTSA set out "to provide a consistent crash problem definition for developers of crash avoidance technologies". They probably had ADAS countermeasures in mind, rather than V2X use cases. As influences converge to cause a crash, aspects of the pre-crash narrative remain in the past. Countermeasures live in the imminent crash moment where the evidence very soon speaks for itself; use cases reach further upstream where experience and imagination are needed to fill in knowledge gaps.

Technological countermeasures have continued apace and positive experience with voluntary application can lead to federal mandates. On June 13, 2023, the *Federal Register* published notice of a proposed new motor vehicle safety standard entitled AEB Systems for Light Vehicles. This vehicle safety system operates at the "crash-imminent" level and combines warning the driver, supplementing the driver's braking and intervening to brake the vehicle. The objective is to prevent the crash entirely, or at least reduce its severity if prevention is not an option. AEB emerged following some success with more limited versions made available as ADAS, including FCW, crash-imminent braking (CIB) and dynamic brake support (DBS). Such systems were voluntarily made available at the discretion of a number of automakers during the 2010s and fall under the automakers' broad rubric of ADAS.

Note that AEB is still considered to be a form of ADAS and therefore engineered by Tier 1 suppliers and offered by automakers, who also carry out the vehicle integration. The proposed rulemaking includes PAEB, a sharpened version of AEB that needs to be tested specifically for pedestrian avoidance. This illustrates the active progression of ADAS systems and the blurred line between ADAS – which is an optional form of driver assistance – and mandatory intervention.

ADAS is far from obsolete and still needs to be seriously considered. But how is it possible for automotive consumers to understand subtle distinctions between cyber functions? The difficulty may be seen in such distinctions not being fully appreciated even by insiders and experts. So, several prominent automotive industry bodies, including the Society of Automotive Engineering (SAE) International, and the American Automobile Association (AAA), came together to clarify ADAS in the mind of consumers (12). Their actions recognized that the brand-specific influence of automotive marketing often leads to a labeled packaging of ADAS that may obscure the individual use cases.

Numerous ADAS applications are identified in the industry document and are listed under several functional headings, namely:

- Collision Warning;
- Collision Intervention;
- Driving Control Assistance; and
- Parking Assistance.

Note that, even within the technology class of ADAS, we are seeing a functional spectrum from assistance, to warning, to intervention. The over-arching characteristic of ADAS remains helping the driver, not replacing the driver. It is instructive to consider the important case of AEB. AEB, included by the industry as a form of ADAS even though it is about to enter Federal Motor Vehicle Safety Standards (FMVSSs), is listed as a function of Collision Intervention.

As AEB enters the world of federal rulemaking, it is no longer merely an optional ADAS function. NHTSA's Notice of Proposed Rulemaking (NPRM) (13) makes reference to a federal program of "safety interventions to prevent crashes" of which AEB is one. Although the FMVSSs have long spoken broadly of two groups of standards – passive safety and active safety – technological crash avoidance standards have been few and far between. The most famous example is FMVSS No. 126 Electronic Stability Control (ESC) Systems. This highly effective standard was phased in for all new light passenger vehicles, beginning in the model year 2009. The advent of AEB as a new FMVSS is therefore an event of significance in all of road safety.

As part of the FMVSS, AEB is no longer a driver assistance use case. It becomes a national crash-prevention intervention. As part of the federal curation required, AEB is a fully defined automotive product subject to a performance specification, component definitions and technical disclosures. Test methodologies are specified along with a battery of tests with pass/fail criteria. Protection against system malfunction is required. Unlike rule 126 for ESC, which is largely invisible to the driver, the AEB intervention interacts with the driver up to a point before the machine takes over. So the AEB rule opens up new territory in motor vehicle safety regulation: it is a phased combination of alerting the driver and bypassing the driver.

5.7.2 THE NEW LANDSCAPE OF TECHNOLOGICAL INTERVENTION

The broad field of technological support and intervention ranges across several important dimensions affecting safety. These dimensions all radiate out from the driver and speak to their human factors responsibilities. Those responsibilities serve to penetrate the interface between the human and their cyber-physical environment as it exists while driving. Human factors engineering has long been road safety's silent partner. How can the machine better suit the human? And how could the system – with the human in the loop – perform better?

David Shinar's important book *Psychology on the Road: The Human Factor in Traffic Safety* (14) provided the foundation for modeling how humans drive and therefore for improving the process using insights from applied psychology. He proposed a hierarchical model of several layers:

- The operational or control level – quick responses to a changing environment;
- The tactical level – mastering of traffic situations;
- The strategic level – where and when am I going?

Building on these ideas, other researchers in the benevolent field of human factors (15) have added two dimensions to Shinar's model:

- Information processing – with stages of perception, processing and action; and
- Task performance – knowledge-based, rule-based and skill-based.

Our field of technological intervention concentrates on certain parts of this conceptual driving model. Tasks at the tactical level are an important part, dealing with the environment – including traffic signals – and interacting with other road users, such as yielding. Tasks at the control level are much more frequent, "almost continuous" (15). Researchers distinguish between these task levels on the basis of complexity and primacy. Tactical traffic tasks are deemed more complex, and control tasks more pressing. Control tasks are also reliant on the driver's acquired skills and tend to be "conducted automatically and very quickly". Relative to this simplified model of driving tasks, much of our driving assistance technology is concentrated mainly at the tactical, or *maneuvering level*, and also at the *control level*.

These are the main ergonomic levels of driving. This scientific term is used to indicate that the driver is required to exercise information processing and to take action as a major part of the system of driving. These tasks place the driver firmly in the framework of ergonomics or human factors, the appropriate discipline for understanding how well they discharge their responsibilities. These responsibilities arise initially from the driver's most fundamental human factors relationships. The first occurs in their response to the demands of the road system; the second arises in their accommodation of the actions of other drivers. A third responsibility is expressed through the driver's control of the vehicle. In order to be safe, the driver needs to take good care of all three of these relationships. That is, they need to respond to the demands of the road system, accommodate other drivers and control the vehicle.

5.7.2.1 Ergonomic Relationship #1

The demands of the road system equate primarily to traveling at the decreed speed, stopping at red lights, negotiating various kinds of junctions and remaining in designated traffic lanes. The response of the driver requires informed attention over both short and long-time scales. Response is also predicated upon habits that are acquired and accepted through driving experience. To improve this relationship, drivers need more actionable information to assist their decision-making. Such information would be pre-processed, in a regular flow, rather than sporadic emergencies. It is incumbent upon roadway operators – those who create the demands – to find ways to convert the demands of the road system into processed, timely information.

5.7.2.2 Ergonomic Relationship #2

Accommodating the actions of others requires drivers to maintain a following distance, accept gaps, adjust speed to a platoon or stream of vehicles, allow for lane changes and allow for the length of queues. This relationship benefits from data that reduces uncertainty and supports a flow of actions. Processing such data is challenging because the actions of other drivers are certain to be highly variable and cannot be predicted. To be effective, the act of accommodation is preceded by a certain element predicting what other drivers may decide to do.

5.7.2.3 Ergonomic Relationship #3

Control of the vehicle speaks to acceleration, braking, lane-keeping, speed modulation, negotiating curves and avoidance maneuvers. Dynamic skills are needed. This relationship benefits from alerts, warnings and interventions, sometimes in emergency mode. Automakers have become extremely adept in enhancing vehicle control and have maintained a continuous transformation of the driving experience at this level.

5.7.3 Distinguishing Safety Interventions

It is not easy to make firm distinctions between the many technological safety interventions we have seen and continue to see. Is there a logical developmental progression? Do some interventions become obsolete when new interventions come along? Or do we need to retain a host of them in parallel? Taking our lead from the ergonomic relationships discussed above, safety interventions are contrived to assist drivers to discharge their prime responsibilities in

1. Responding to the demands of the system of roadways;
2. Accommodating the actions of others using the road; and
3. Controlling their vehicle.

In a generic sense, safety interventions that align with these responsibilities may be described in the following categories:

1. Processed demands – the requirements imposed by the roadway are provided more reliably, clearly, deeply and in a more timely manner; V2I initiated by the roads sector is necessary; associated V2V initiated by the vehicles sector is well-suited;
2. Informed interactions – accommodations for other road users are based on processing their movements; V2V initiated by the vehicles sector and ADAS

offered with vehicles are required; more advanced data processing is needed and may be at hand in the age of artificial intelligence (AI);

3. Dynamic skills – control of the vehicle covers the gamut from assistance to intervention, both overt and covert; ADAS offered with vehicles and FMVSS for crash avoidance are required.

5.8 FROM VOLUNTARY TO MANDATORY APPLICATION

The question of voluntary use of safety technologies versus the judicious use of coercion is ever-present and subject to national mores, ultimately traceable to the philosophies of the social contract, written or implied. Currently, all countries mandate some level of dynamic machine skills in the control of the vehicle; these mandates fall on the shoulders of the OEMs.

Beyond the norms of road rules and traffic control, the demands of the roadway system are not processed or transmitted. Traffic control OEMs are not subject to mandates beyond standardization and quality of traffic control products. There is no equivalent of ADAS in the roads sector, although some degree of informal innovation occurs in the industry. The ITS industry markets its products to the road agencies in an atmosphere of economic constraint and risk aversion. It is therefore difficult for their offerings to rise to the level of recognized safety functionality, as is the case with ADAS.

At a time of road safety stagnation and signs of decline, technological safety interventions are essential for drivers to succeed in their most important ergonomic responsibilities. A paradigm of roadway-led V2I is overdue and currently presents itself as the foundation of future road safety. Much greater clarity is needed; V2I, V2V and indeed V2X should not be presented as ambit claims. V2I has revealed itself as a conduit between the demands of the road system and the response of the driver. V2V is an instrument of accommodation between the driver and the actions of proximate vehicles. ADAS extends the range of the driver's ability to control the vehicle and avoid conflicts. Considering the current point of departure – people doing all of the driving – this technological path is true and will remain true. We have not yet entertained machine driving, but the connected paradigm is also conducive to a future of AVs.

5.9 THE UNDERLYING BELIEF SYSTEM OF V2X

V2X is not a safety solution; it is a paradigm – a model with technical elements bound by a set of beliefs. Such beliefs have not been widely articulated but may be deduced by observing the developmental history of V2X. Right from the early days of VII, there has existed an uneasy dependency between the constituencies of roadway systems and vehicles. Not only do they speak a different language and develop on very different trajectories, but industries have a basic tenet of not relying on the government. And the sensitive nature of VII mobility data – raising the specter of tracking and monitoring – caused worries about government access. Just who would be permitted access? The industries involved shied away from such discussions with the government. And the government was nervous about deploying a certain wireless technology standard when the too-nimble private companies could make a rapid business decision to change methods. This divide prevented a lot of important discussions from happening.

So, with the benefit of hindsight, what are the aspirations of V2X, and what are its limitations? And what do we not know about V2X that we should know? V2X is imbued with the human factors ethos of making driving situations as tractable as possible for humans and also making driving as safe as possible. The freedom and anonymity of driving cannot be sacrificed, and errors cannot be punished. V2X is the expression of the human factors idea of driving as a serious undertaking requiring foresight, preview and informed attention in the moment. V2X proceeds on the assumption that there is a core set of responsibilities that underlines all of driving; any and all deficiencies in their discharge increase driving risk. All of the other things that drivers need to do, such as navigation, are not the business of V2X. V2X does not supersede or replace other safety technologies such as ADAS and FMVSS for crash avoidance. To summarize, *V2X is driver-facing situation awareness in a full human factors framework*. This framework covers all of the interactive responsibilities of drivers within the system of road traffic.

Certain other principles are important to discuss and are not necessarily agreed. All sectors of the road transportation system should collaborate and contribute. The history of V2X has seen wild swings in first mover responsibility; we started with the infrastructure, then gratefully handed the baton to the automotive sector, and now we are back to the infrastructure franchise. It now makes more sense than ever that the first actions to disseminate connected vehicle technology should come from the road sector. The automotive sector and their customers need to have confidence in technically sufficient wireless located strategically in the road network. Only then would vehicles carry conforming situation awareness technology.

Because V2X came into being principally as a breeder of safety countermeasures, we lack concise, instantaneous ways to convey situation awareness. Many questions need to be answered concerning "how much information, how often?". One point of discomfort in the Ann Arbor FOT was not being sure which other vehicle an alert was referring to. Can such information be conveyed in a non-distracting manner? Can more prescient situation awareness reduce the frequency of required lane changes? Are there baseline safety maxims – responsive to prevailing roadway and traffic conditions – that could be conveyed in a calm, consistent manner? For example, "stay in your lane for the next 3 miles".

Part of the reason for the current V2X hiatus in deployment stems from the safety enterprise's faith in countermeasures. The ability of safety professionals to deliver reductions in fatalities and serious injuries depends on specificity, and ways have been found to reverse-engineer countermeasures. Much of the high status of road safety comes from its focus on human harm, and demonstrable reductions in that harm. As we have seen, V2X is not necessarily in the business of countermeasures. The paradigm of V2X is breaking important new ground by addressing the essential ergonomic relationships entered into by all drivers as soon as they take the wheel. V2X provides targeted assistance across the board: responding to the demands of the road system, accommodating the actions of other drivers and controlling the vehicle.

V2X does not currently provide all the answers when a crash is imminent. But the ergonomic tonic of situation awareness is needed every day – every hour, every minute while driving – and that is what V2X offers. In terms of driving ergonomics, roadway intersections are an awareness desert. One simple example is the non-signalized intersection fitted with a stop sign. Having stopped, the driver has no idea when to proceed across the intersection. This hazardous scenario occurs repeatedly across the country, every day. And it leads to one of the most frequent and injurious types of collisions. In the language of V2X, the vehicle that stops is the "host vehicle" and the vehicle that crosses is the "target vehicle". Good ergonomic practice says that the stationary vehicle needs to "host" situation awareness data, perhaps pared down to alerts about when not to proceed. But the vehicle in motion also needs to be a host – to know that they are not going to be a "target" and are clear to proceed. Such wireless exchanges are not occasional countermeasures but are a constant force for good under the paradigm of V2X.

5.10 V2X IN OTHER COUNTRIES

Reflecting the global sharing of ITS research and development, the paradigm of connected vehicles and infrastructure has received major attention in regions and countries around the world. A multitude of collaborative efforts blossomed, especially during the decade of the 2010s. While certain countries including the United States, Australia and Japan concentrated on the safety benefits of the new paradigm, Europe embraced a more ecumenical approach to include traffic efficiency and vehicle emissions as well as safety.

It is no surprise that each country or region decided to stage their own demonstrations and model deployments: while V2X is highly technical, it is also a major exercise in public–private, cross-sectoral collaboration. It is by nature much more local than the auto industry. It is also subject to influences of the physical environment that affect wireless transmissions. And it needs to be tested under a range of climatic regimes. Business models are not well established and may prove to be variable across countries and regions.

The use of V2X for the specific purpose of crash avoidance was always a central goal in the United States. However, a more general set of applications were included elsewhere. Some of the use cases considered may be described as safety-related – such as traffic jam advisories – but do not rise to the level of assisting drivers to avoid crashes. Many of the use cases amount to general driver assistance that may be well removed – upstream – from crash causation. The approach taken by the European Commission (EC) (16) was widely emulated in other countries, including the use of the term C-ITS.

The EC carried out large economic studies that created alternative scenarios for the rollout of C-ITS. These studies were reviewed in 2016 (16) and broad economic analysis was carried out on several scenarios of European deployment. The analysis adopted a large number of diverse use cases which were allocated in bundles.

Calculation of economic benefits needed estimates of the effectiveness of each use case in influencing driver behavior. Typically, each use case would be assigned a low and high estimate. Such numbers may pass the test of reasonableness but are not based on research evidence. The scenarios were created with various combinations and timing of the bundles, applicability to different road classes, plus regions and corridors of deployment. So-called Day 1 C-ITS services included some hard V2X crash avoidance use cases, as included in the early days of U.S. VII, and continued through to the Ann Arbor Safety Pilot FOT.

This EC study provided an expansive listing of C-ITS use cases and described them in terms of likely technology (V2I or V2V), location (urban or rural) and wireless regime, among other factors. From the ergonomic point of view of a driver, the use cases may be classified as follows:

1. Driver responding to the demands of the road system
   a. In-vehicle signage (V2I)
   b. In-vehicle speed (IVS) limits (V2I)
   c. Roadworks warning (RWW) (V2I)
   d. Green Light Optimal Speed Advisory (GLOSA)/Time to Green (TTG) (V2I)
   e. Wrong-way driving (V2I)
   f. Signal violation/intersection safety (V2I)
2. Driver accommodating the actions of other drivers
   a. Emergency vehicle approaching (V2V)
   b. Slow or stationary vehicle (V2V)
   c. Traffic jam ahead warning (V2V)
   d. Shockwave damping (V2I)
   e. Motorcycle approaching indication (V2V)
3. Driver controlling the vehicle
   a. Emergency brake light (V2V)
   b. Vulnerable road user protection (pedestrians and cyclists) (V2I and V2V)
   c. Cooperative collision risk warning (V2V)

Note that other use cases considered by the EC under the rubric of C-ITS are not included in this classification because they are not interactive responsibilities requiring situation awareness. Examples include hazardous location notification, probe vehicle data, weather conditions, traffic signal priority request by designated vehicles, information on parking and fueling, traffic information and smart routing and zone access control management. Overall the EC use cases illustrate a focus on the demands of the road system and accommodations for other drivers. Their framing of C-ITS is less focused on emergencies and imminent collisions.

Major European consortia and deployments included the Car2Car Consortium, which helped to secure the necessary wireless bandwidth, drove important standards and contributed to the validation of V2V. In another example, the Amsterdam Group set up a C-ITS Corridor across Austria, Germany and the Netherlands using V2I radios located in work trailers at roadwork sites. This effort included two applications: roadworks warnings and probe vehicle data. These test deployments, and many other projects, were usually concerned with mobilizing the necessary collaborations and testing institutional frameworks rather than providing a solid picture of safety benefits. The economic studies had already led governments to believe in the over-arching societal benefit of C-ITS. But the evidence was not sufficiently compelling for them to become the first mover in C-ITS deployment at scale: reputational and financial risks underscored a cautious approach.

The Australian State of Queensland launched a substantial C-ITS safety test program in 2019, known as the Connected and Automated Vehicle Initiative (CAVI). Carried out under CAVI, the Ipswich Connected Vehicle Pilot (ICVP) (17) included 355 participants and 29 RSUs. Driver warnings included red lights, roadworks, road hazards and the presence of pedestrians. Queensland's Transport and Main Roads (TMR) concluded that the delivery of warnings has a positive impact on driver behavior. The success of CAVI encouraged the federal authorities to seriously entertain C-ITS as a safety solution. However, Australian states have operational authority for road safety and are not yet seeing sufficient evidence to create a major shift in road safety policy.

The TMR report describes six V2I use cases included in their trial, as follows:

- Advanced Red Light Warning (ARLW)
- Turning Warning Vulnerable Road Users (TWVR)
- Road Hazard Warning (RHW)
- Back-of-queue (BoQ)
- RWW
- IVS.

Almost all of these use cases equate to use cases cited in European C-ITS, but with some local variations in terminology and perhaps intent. TMR's objectives for the trial included safety validation but also extended to demonstration of the technologies, growing institutional readiness and building public and private partnerships.

In 2021, the Toyota Motor Corporation carried out a controlled C-ITS project (18) in Melbourne, Victoria, using the inner urban test bed known as the Australian Integrated Multimodal EcoSystem (AIMES). AIMES is operated by the University of Melbourne in partnership with the Victorian Department of Transport and Planning (DTP) and others. Use cases verified included

- ARLW
- Turn Warning Vulnerable Road User
- RHW
- BoQ warning
- RWW
- IVS.

These safety applications – some new, some modified versions of previous use cases – were developed by Toyota, building on their experience with CAVI in

Queensland. The appropriateness of the use cases, and their improved algorithms, benefited from the technical acumen of a major global automaker. Note that the six use cases were derived from the set previously used in Queensland's ICVP but were not necessarily identical in their functionality. It is also noteworthy that these use cases all operate under V2I, even though they were stage-managed by an automotive OEM.

These and many other projects have shown that V2X, or C-ITS, is a lot more than a technology – it is driving ergonomics. It needs to be applied, and it can be applied in many different ways. Strong focus and decision-making in V2X have thus far eluded the many countries and jurisdictions where it has been embraced and tested. While V2X is mentioned in many national safety strategies, it is not yet relied upon. Why? From an engineering viewpoint, we are dealing with system innovation – a demanding activity that needs iteration upon iteration. This often-chaotic process is difficult to sustain in a shifting multi-partner environment. There is also a need to move beyond the universal appearance of V2X and better reveal its safety purpose.

Arguably, the prime purpose of V2I technology is to promulgate the active demands of the road traffic network directly to the driver. V2I is capable of being more demanding than signs and signals and is harder for the driver to ignore. In a similar vein, the prime purpose of V2V technology is to alert the driver concerning the impositions of other drivers. V2V does this by far exceeding the driver's field of view and processing multiple targets in parallel. It is time for the highway sector to unequivocally embrace V2I technology in pursuit of 21st-century safety innovation.

REFERENCES

1. Resendes, Raymond (2005) Vehicle Infrastructure Integration. SAE Government Industry Meeting. Washington, DC. May 10, 2005.
2. Hollnagel, Erik., Wears, Robert L. & Braithwaite, Jeffrey (2015) From Safety-I to Safety-II: A White Paper. The Resilient Health Care Net.
3. McCall, Janice R. (2020) Knowing versus Understanding: Adjusting the Contextual Lens in Safety Science. MATEC Web of Conferences 314, 01008 (2020).
4. Piaget, J. (1973) *To Understand Is to Invent: The Future of Education*. 53. Grossman Publishers.
5. United States Department of Transportation (2022) National Roadway Safety Strategy. Version 1.1. January 2022.
6. United States Department of Transportation (2022) Saving Lives with Connectivity: A Plan to Accelerate X2X Deployment.
7. United States Department of Transportation (2015) Safety Pilot Model Deployment. Lessons Learned and Recommendations for Future Connected Vehicle Activities. Final Report – September 2015. FHWA-JPO-16-363.
8. National Highway Traffic Safety Administration (NHTSA) (2011) Integrated Vehicle-Based Safety Systems (IVBSS) Light Vehicle Field Operational Test Independent Evaluation. DOT HS 811 516. October 2011.
9. National Highway Traffic Safety Administration (2007) Pre-Crash Scenario Typology for Crash Avoidance Research. DOT-VNTSC-NHTSA-06-02.

10. National Highway Traffic Safety Administration (2015) Independent Evaluation of Light-Vehicle Safety Applications Based on Vehicle-to-Vehicle Communications Used in the 2012–2013 Safety Pilot Model Deployment. DOT HS 812 222.

11. United States Department of Transportation (2017) Vehicle-to-Infrastructure (V2I) Program: Research, Development and Deployment Support conducted through 2020. Final Report – December 31, 2021. FHWA-JPO-21-831.

12. Society of Automotive Engineers International et al. (2022) Clearing the Confusion: Common Naming for Advanced Driver Assistance Systems. July 25, 2022.

13. National Highway Traffic Safety Administration (2023) Federal Motor Vehicle Safety Standards: Automatic Emergency Brake Systems for Light Vehicles. Notice of Proposed Rulemaking. 49 CFR Parts 571 and 596 [Docket No. NHTSA-2023-0021]. RIN 2127-AM37.

14. Shinar, David (1978) *Psychology on the Road: The Human Factor in Traffic Safety.* John Wiley & Sons, Inc.

15. Theeuwes, Jan (Ed.) (2017) *Designing Safe Road Systems: A Human Factors Perspective.* Routledge, CRC Press.

16. European Commission (EC) (2016) Study on the Deployment of C-ITS in Europe: Final Report. DG MOVE. MOVE/C.3/No. 2014-795.

17. Queensland Transport and Main Roads Department (2022) Ipswich Connected Vehicle Pilot. Lessons Learned Report. March 2022.

18. Australian Integrated Multimodal EcoSystem (AIMES) (2022) Enabling Infrastructure to Vehicle Communication for Safety Applications of Connected Vehicles in Carlton Victoria – Initial Test. The University of Melbourne. eng.unimelb.edu.au

6 The Leap toward Automated Vehicles

Automated vehicles (AVs) are widely regarded as a future road safety solution of vast importance. It is popularly understood that human error causes the vast majority of crashes and fatalities. It therefore makes sense to many that humans should be replaced by machines that are never distracted, angry or tired. So it is important to consider vehicle automation as an emerging safety solution, of a highly technological nature, as well as the ability of automation to take up that enormous mantle without creating other safety problems. For example, are AVs going to be able to drive well enough for the many miles between their infrequent episodes of avoiding an imminent crash?

The popular notion of an AV safety transformation requires examination from several points of view. Firstly, humans cannot be simply replaced because many will continue to drive for years into the future. Automated driving systems (ADSs) are expected to provide an alternative, but in doing so they introduce a new and unknown element in the driving system. Secondly, human drivers have a very low failure rate; can it be proven that ADS's will consistently do better? Thirdly, will ADSs be able to sustain a new modus operandi, a new steady state version of the dynamic of driving? Humans provide an enormous value in the daily fabric of driving; will ADSs be able to emulate that, decade after decade? How will we navigate the transition from the old status quo to the new situation?

Undoubtedly, AVs are an irresistible force and may prove to be the most impactful system innovation of our era. Meanwhile, technology has become a much bigger influence on road safety in the new century. The gap between vehicle engineering and road engineering has widened and we look more to automotive manufacturers and their technology suppliers to come up with safety solutions. However, the idea of cars that drive themselves and are capable of replacing humans on a broad scale requires a certain leap of the imagination.

That is a leap worth taking. Most developers of AVs have responded with innovation of a quality and scale rarely seen in the automotive world. The addressable market for AVs is sufficiently large to inspire the technology and communications sectors as well as the automotive sector. Coincidentally, the challenge of mobile data processing has transformed the computer industry. At least some AVs are probably good enough to take on the ergonomic roles of the human driver in substantial

DOI: 10.1201/9781003483861-6

environments. But how do we get from there to the majority of drivers, who represent a massive quantity of driving, being traded in for machines? What steps would we need to take to bring about such a quiet revolution? And is that really an ethical goal?

6.1 THE CHALLENGE OF MACHINE DRIVING

All previous safety technologies have been developed with a focus on harm reduction. In contrast, driving machines have been invented in the positive – in living color – with the goal of driving better than humans can. Driving machines are not intended to be optional or buried under the hood – they are meant to be out there in the light of day for all to see. They do not represent a variation or tweak to the current system of driving – they change it dramatically and it is no longer the same system. The automotive technology content increases by several orders of magnitude. AVs utilize a range of sensing technologies and maps to create situation awareness, software and actuators to provide control functions (steering, accelerating and decelerating) and driver interfaces. Each AV hosts significant computational and data storage capabilities. The challenge does not end there. AVs may also use internet connections to host less time-critical data, or to download relevant software, and software updates. Such links may also provide manufacturers with powerful R&D information as they develop self-driving features from the "beta" to the commercial stage.

High levels of automation open the door to radical changes in the architecture and design of vehicles. As the purchasers of automobiles become less wedded to requiring the "ultimate driving machine", dramatic changes in powertrains, size, weight, seating arrangements, amenities and cargo capacity may be realized. There is a wide range of levels of automation, as companies from the technology sector and automotive industry vie to create the automated product. Concepts could range from "everything somewhere" – totally self-driving but suited only to certain environments – to "something everywhere" where worthwhile but limited automation may be used quite widely (for example, while traveling on major highways).

So far, governments have played a limited role for AVs. A number of U.S. states have adopted legislation applying to the testing and operation of AVs. There may be requirements for licensing, incident reporting and human occupants, and limitations on driverless operation. However, many states have no such legislation, although freedom to operate is not clear. Many expect federal safety rules to eventually apply to AV performance. Standards development is continuing on several fronts, and it is likely that standards will operate on at least two levels: (1) the more traditional performance specification where physical testing is implied, and (2) safe systems, functional safety and software design standards. The development of norms for human intervention with the machine, and for the coexistence of automated and traditional vehicles on the same roadways, is considered by many to be a significant challenge.

The hard work of proving the feasibility of AVs operating in and around traffic has proceeded throughout the 2010s. Such vehicles have been, and continue to be, used in vocations such as ride hailing in cities and in over-the-road freight delivery. In these cases, operational cost reduction is more important than safety improvement, but such applications do provide certain proof of safety at scale. But automation as a widespread national mobility solution remains daunted by major national preoccupations such as the manufacturers' responsibility for delivering a safe vehicle, liability and cybersecurity. Beyond the potential importation of criminal cyber behavior into the family car, hackers' malevolent intentions to disrupt the national transportation system should not be discounted.

Non-traditional manufacturers of AVs are oriented to new mobility services, such as low-speed driverless shuttles in cities, or on restricted campuses (like universities, hospitals, business parks, airports and shopping complexes). Such limited deployments have served to raise awareness of AVs; they do not pose significant safety issues in themselves but do not address societal interests in the reduction of motor vehicle harm. Another limited early adopter of technology akin to AVs is the platooning of freight trucks. This is not driving automation *per se* because the human driver in the lead vehicle still does all the driving. Platooning relies on wireless communication between the lead vehicle and one or two following vehicles that have dormant humans in the drivers' seats; it is as much a connected vehicle application as it is an AV application.

There is a big difference between media interest, coverage of limited applications and public approval. Most of the drivers and passengers who rely on daily use of motor vehicles for essential life purposes have not experienced AVs, nor are they versed in the narratives underlying their development. A comprehensive 2021 review (1) of international public opinion studies found deep-seated concerns about the safety of AVs. These levels of concern have been linked to publicity about AV crashes and to discourse about ethical dilemmas on the part of driving machines: are they programmed to follow preferred crash scenarios that reduce harm to selected types of "other" road users? When contemplating purchase of an AV, potential customers want to be shielded from liability and are generally reluctant to pay more for an AV.

But much depends on the unspoken assumptions of the questions being asked. Periodic surveys carried out by the British Institution of Mechanical Engineers (2) show that opinions on AVs are well divided; but majorities have concerns about AVs sharing the roads. They are a little happier with AVs that retain standard driving controls but are opposed to those with no human control. An Australian study published in 2019 (3) showed a wide range of AV acceptance, but generally the Australian public approved of the expected benefits of AVs including safety; they were also more willing to ride in AVs than potential users in other countries. However, the Australians also reported a low level of knowledge about the operation of AVs. It may be that AV acceptance on the part of governments – and even extending to promotion of AVs for their safety benefits – has already influenced users, at least in Australia.

In 2022, the AAA Foundation for Traffic Safety published the results of periodic sample surveys of the U.S. population and their attitudes to AVs (4). These surveys were framed in a knowledgeable way that included capability levels, forms of unsafe driving and sources of user concern. A significant majority was aware of the capability levels of AVs and between one third and one half were confident about AVs' ability to prevent crashes. Almost 60% were confident about AV crash prevention in the event of driver lapses including distraction, drowsiness and impairment. But the quality of the technology is still an important issue in the minds of consumers: at the higher levels of AV capability, potential users cited technology malfunction as their biggest concern with AVs.

6.2 AV LEVELS OF CAPABILITY AND OPERATING ENVIRONMENTS

The six levels of driving automation defined in Society of Automotive Engineers (SAE) Standard J3016 (5) have been widely adopted around the world. The essential distinctions made are as follows:

- Level 0 – warnings and momentary assistance as in ADAS applications
- Level 1 – machine supports driving through either steering or braking
- Level 2 – machine supports driving through both steering and braking
- Level 3 – machine driving in limited conditions with occasional support
- Level 4 – machine driving in limited conditions
- Level 5 – machine driving.

The human drives the vehicle at all times under Level 0 through Level 2; the ADS drives the vehicle at all times under Level 3 through Level 5. The standard refers to the Dynamic Driving Task (DDT), the most immediate part of a driver's responsibilities. The specified level of the DDT applies only within the Operational Design Domain (ODD) specified by the manufacturer. This specification is provided at the manufacturer's discretion and may include the operating environment, restrictions of geography, time of day or weather and stipulations of traffic or roadway characteristics. If the vehicle exceeds the boundaries of its ODD so nominated, the ADS automatically returns to a minimal safe condition.

Note that these levels only apply to the host vehicle, and no account of other proximate vehicles is taken. Consideration of other vehicles only comes in peripherally through the ODD. For example, the ODD could exclude heavy traffic; in the alternative the ODD could call for certain high traffic conditions that would also imply lower speeds.

Much has been made of these levels and their numerical escalation. In practice, Level 4 is the highest form of driving automation. It is difficult to imagine a manufacturer offering a totally unbounded Level 5 vehicle. Most of the dialog comprising the annals of AVs centers on Levels 3 and 4. Level 3 has attracted controversy because it requires drivers to immediately resume control of the vehicle, with no specific lead time or warning. Human factors principles do not support sudden cognitive demands being made on dormant drivers who have suspended their situation awareness. Nevertheless, Level 3 has survived and prospered and is finding its way into certain models manufactured by global Original Equipment Manufacturers (OEMs), such as Mercedes-Benz. On the other hand, Level 4 is required in driverless urban mobility services such as those operated by Waymo and GM Cruise. If this bifurcation continues, there will be a continuing demand for Level 3 in privately owned automobiles and for Level 4 in ride service fleets.

On January 26, 2023 Mercedes-Benz USA (MBUSA) announced their intention to market the availability of Level 3 "conditionally automated driving" in their

S Class line of the 2024 model year (6). This automated driving solution – comprising strategy, rules and integrated technology – is called Drive Pilot. These AVs first entered the U.S. market in the State of Nevada after being certified under Nevada Chapter 482A for Autonomous Vehicles (7). An important part of the Drive Pilot solution is use in traffic of higher density that is traveling at low-to-moderate speeds up to 40 mph. The driver is alerted when the system is available and may choose to activate the automated mode. Additional aspects of the Mercedes-in-Nevada Level 3 solution include smoothing of the handover from automated driving back to human control and reversion to a minimal risk condition in the event of malfunction.

Level 4 Waymo operations in the State of California are expected to comply with a Statement and Map of ODD – Driverless Deployment (8) that was approved by the California Department of Motor Vehicles on November 9, 2022. The Waymo-in-California Level 4 solution covers most road types and applies to posted speed limits up to 65 mph. It allows Waymo autonomous passenger vehicles to transport both people and goods for reward. Waymo's dynamic operations can change domain constraints at any time; they can also limit commercial operation in response to certain roadway or climatic conditions. The geographic ODD is specific to the City of San Francisco and a portion of San Mateo County.

Stark differences have emerged between AVs at Level 3 and Level 4. Because Level 3 vehicles need to return control to the driver from time to time, they need to be conventional from the perspective of driving position and controls. They must provide adequate field of view and meet all of the federal motor vehicle requirements, including passive safety. Ultimately, they remain consumer products destined for private ownership. On the other hand, Level 4 vehicles are likely destined for commercial purposes that operate outside the traditional axis between OEMs and their customer base. So far these operations are tightly constrained and difficult to generalize. The idea of a company such as Waymo being remotely responsible for "dynamic operations" is a new element in the AV solution. And we have already learned that levels of automation and statements of operating environment are not sufficient to define and therefore deploy an AV.

6.3 COULD AVs ELIMINATE HARM IN ROAD TRANSPORTATION?

Government transportation agencies have an intense interest in the potential for AVs to eliminate human driving error, which experts say is the principal cause of all crashes. This will require the widespread use of quality AVs; it will represent not just a change in road safety – an additional string to its bow – but a change in the road transportation system itself. This is clearly a weighty matter. While public sector interest in a new and powerful safety weapon is high, not everyone in the road safety fraternity is totally on board with automation. Quite rightly, there are objections about mixing AVs with traditionally driven vehicles, about certification of AVs for use on public roads, and many, many more questions are being asked. Some of the questions being raised require us to compare the decisions of machines with those of people, when we know so little about what people are thinking when they drive.

Inside road safety, there may also be a subtle objection in that going all in on AVs may divert resources from existing road safety programs within safety's comfort zone of countermeasures. Inspection of current national road safety strategies reveals little sign of reliance on AVs; there is a reluctance to place too much weight on such disruptive technology.

On the face of it, good AVs would reduce physical harm because the first (physical) crash would not occur. Good AVs could be programmed to deal with the major crash types such as drifting out of the lane, rear-ending stationary vehicles and failing to account for cross-traffic at unsignalized intersections. In doing so, good AVs would also remove certain mental harm of human lapses and failures. AVs should help to promote good mental hygiene in driving; they need to be unequivocal about situations, options and decisions, and need to act firmly and instantaneously.

6.3.1   PROGRESS WITH DRIVERLESS TECHNOLOGY

Much progress has been made in creating the technology of AVs. On March 13, 2004, the U.S. Defense Advanced Research Projects Agency (DARPA) launched a robotic vehicle race in California's Mojave Desert. This has been widely reported and became the subject of many books; for example, former R&D supremo at General Motors Larry Burns recounts the successful GM team he built with Chris Urmson and his colleagues from Carnegie Mellon (9). The first prize of 1 million dollars was not claimed, because no one finished a rough and demanding course of 150 miles. This, the very first DARPA Grand Challenge, planted the seeds of change across the transportation industry. It was repeated the following year and five vehicles completed a desert route of 132 miles. Subsequently, there were more and different grand challenges, including an urban grand challenge. Over the years, hundreds of teams have competed, entering robotic vehicles with names like Sandstorm, Boss and Junior.

The challenge had ignited a competitive effort to develop driverless cars. By 2007, major research universities such as Stanford, Carnegie Mellon and MIT had teamed with global automakers, including General Motors and Volkswagen. After many years of niche interest on the part of inventors and entrepreneurs all around the world, driverless cars had become a bona fide R&D project with commercial interest beyond mere curiosity. The Californian teams were particularly energized and the first industrial-scale effort to develop a practical self-driving car quietly commenced on January 17, 2009. Tellingly, this enterprise was not launched by an automotive OEM. The project was housed in a Google X lab in Mountain View, California. This technology was envisioned as universal, rather than being subject to the complications of restrictions: caveats on the use of the Google car were to be avoided.

Suffice it to say that the project is continuing undiminished, now known as Waymo. It is still controlled and funded by Google – restructured as Alphabet. Almost 15 years later, the project has proven to be more than a grand challenge. Unlike automation in planes or trains, automobiles operate in environments of such complexity that it is almost impossible to decide when enough R&D is indeed enough. Waymo, and other corporations like them, have to decide if and when their ADS is safe for general use on public roads.

The National Highway Traffic Safety Administration (NHTSA) has an intense interest in the safety of AVs and has put forward guidelines (10) for their design and testing. But NHTSA remains very far from making hard pronouncements on the suitability of AVs for general use. For example, they advise manufacturers to consider placing conditions on where and when their ADS is designed to be used – but this is at the discretion of the manufacturer.

At the same time, NHTSA and government safety agencies all over the world have spoken loudly and consistently about the tremendous road safety potential of AVs. Sixty years of continuous learning in safety have brought us to a place where "driver error" stands before us as a mountain to be climbed. If drivers, with their many potential flaws, could be replaced by steady, accurate, predictable, reliable machines then that mountain would be much less formidable; some would even say it could be leveled. It is therefore ironic in the extreme to be held back by uncertainty about safety. We need to do better than a statement like "automated driving systems offer unparalleled road safety potential, if only we could be assured that they are safe".

6.3.2 FROM CRASH AVOIDANCE TO "SAFE PASSAGE"

The road safety we know has been based on the study of crashes, while safe AVs are not supposed to crash at all. Apart from anything else, we are flipping safety's ethical underpinnings from consequentialism to good will and rules – sometimes called morality and deontology. That is, our focus is veering from outcomes, including injuries and fatalities, to a more rules-based system that always keeps vehicles apart. It would be difficult to over-estimate how wedded we have become to road safety consequentialism. Fortunately, the information being provided to NHTSA concerning the safety design principles of certain individual companies indicates that there is indeed a new way to think about the AV safety problem.

Let's call it "safe passage", where the AV is able to proceed to its destination in a purposeful manner, all the while using its braking, steering and acceleration controls to totally avoid contact with other vehicles, infrastructure, roadside furniture and objects. These other vehicles could be AVs or conventional vehicles with human drivers. Perhaps using different methods, we would also include avoidance of pedestrians, cyclists, scooters and the like. Safe passage might be visualized more as a video game rather than a slow-motion replay of a bobbling crash test dummy. It is no accident that many of the computer chips used in driving machines are made by NVIDIA, who cut their teeth on chips for video games.

An interesting language is being invented by companies involved in proving safe passage via the actions of an ADS. In the case of leading chipmaker NVIDIA (11), a relative newcomer to the automotive world, all vehicles that are relevant to safe passage of the protagonist AV are called moving actors. All moving actors have a "safety procedure" – the minimum control actions to guarantee no contact with other actors. They also have a "claimed set", comprising the safety procedure, plus a potential expansion of the vehicle's envelope in the immediate future. This expansion is defined by maximum control actions. The "minimum symmetric responsibility" refers to the control actions by all actors that just guarantees zero contact between vehicles. It would also be reasonable to ask for greater control action to allow for irresponsible actors. The expectation is that all actors are required to help avoid encroachment of "claimed sets". The "safety force field" embodies the contention that occurs between actors when they try to claim the same parts of space and time. The force field tells the ADS the most productive type and direction of turning, deceleration or acceleration. When yielding is necessary, the principle is: right of way is given, not taken. This leads to the idea of a "safety cocoon" that is always on. A collision risk is always a collision risk, no matter who is right. In the case of the protagonist insisting on carrying out a lane-change when there is not enough space between vehicles in the adjacent lane, the protagonist is called the "ego vehicle".

This abstract language is needed to visualize how the ADS guarantees safe passage on a consistent basis. These concepts, and many others, are needed to continuously create and maintain the AV's "safety cocoon". This helps to explain why safety has always been defined in the negative, by failures – the opposite of what we actually want. Throughout the history of road safety, it has not been possible to sense and compute the safety cocoon and all its many parts. Maybe, technology is allowing us to break through this barrier. We start to see that a whole range of positive control actions, carried out at exactly the right time, continuously refreshed and absolutely unambiguous and consistent, always minimizing contention, actually *are safety*. Getting very far ahead of ourselves, if different ADS manufacturers used the same constituent elements of control and contention we may be able to develop metrics of how well a given ADS provides control and minimizes contention. This could provide us with metrics that quantify the ability of a given ADS to guarantee safe passage. We could stop trying to do hundreds of millions of miles of testing just so that we could study the crashes, as in old-world safety.

6.3.3 LEARNINGS FROM AV SAFETY REPORTS

In recent years, the NHTSA has requested ADS manufacturers to file safety reports. These reports are intended to provide self-assessments of that company's way of guaranteeing that their ADS can drive safely. More than 25 companies have filed reports. There is no standard template, so the approaches used to design a safe ADS – and to convince NHTSA and many others that they are following NHTSA guidance – are diverse in the extreme. One way to look at this diversity is to see how wedded is each manufacturer to 20th-century norms of road safety, or to what extent they venture into new 21st-century concepts of safety that don't always revert to metrics of crash occurrence and severity.

An interesting effort to straddle these two worlds is contained in the 2018 report Measuring Automated Vehicle Safety published by the RAND Corporation of Santa Monica, California (12). The mentality of the report is very much the current era of AV development, demonstration and deployment. It talks about AV safety measures that could be leading indicators, or alternatively lagging indicators. Leading measures include infractions, roadmanship and disengagements of the ADS. Infractions and disengagements are really old-world measures that pertain only to our vehicle, in that they measure machine lapses and failures. But the idea of *roadmanship* takes us to the realm of vehicle interactions and how well they are handled by the ADS, or by multiple ADSs. This leads researchers to the idea of a safety envelope that could be violated by our vehicle, or other vehicles, the issue of who initiated the problem, and also who should respond. Other simpler measures include the conflict metric "time to collision"; that is, if vehicle positions and paths continue as they are, how long until a collision occurs?

In the realm of lagging indicators – which is all we have in old-world safety science – ADS companies tend to compare the safety of an ADS with that of human drivers. But there is a desire to leap beyond conventional metrics like fatalities per 100 million miles of travel, to allow for different and changing operating environments that could affect the ADS more than the human. Please recall that the human driver is supposed to be able to deal with anything that happens on the road; this is not necessarily true for ADSs. The resulting retreat from universal metrics creates new measures that speak to specific places and times, or specific types of places. The ODD defined by the manufacturer could even exclude, say, left turns at intersections. It would not be fair to compare an ADS with this ODD with across the board human driving because humans don't do so well in such situations – so human crash rates would need to adjusted downward in order to have a fair comparison. And very different denominators – measures of exposure – may be needed, such as hours of operation, number of licensed operators or owners, or per city block etc. Things go downhill very quickly when we even think about comparing humans with driving machines. Nobody is willing to make such machine versus human comparisons at this point of ADS development and rollout. This is currently the wrong question.

At the present time, ADS manufacturers' safety filings show very different approaches to safety assurance. Waymo (13) is the current market leader, or developmental leader, because there is very little in the way of a commercial market at this point. They are very conscious of defining the ODD but are still striving toward an "anytime, anywhere" tagline. Waymo have combined the ability of the ADS to drive safely "in the moment" – akin to old-world safety – with a new emphasis on the system safety embedded in the vehicle's extensive software. There is great emphasis on the ADS sensors' ability to detect objects and events, and respond, as well as the autonomous driving software. Then, physical and simulation testing is carried out to verify the base vehicle safety, autonomous driving hardware and autonomous vehicle software. In the latter case, Waymo are looking at behavioral competencies for normal driving. While the latter competencies are entering the domain of new-world safety, they are very much tied to tried-and-true crash types and complex, unusual driving situations that challenge the ADS. Waymo seem to be saying that a good ADS must do better than humans in their worst moments, and also must be aware of its own weaknesses. And they are keenly aware that human and machine weaknesses are two different things. Just because a machine does well in an area of human weakness it is not inured against its own weaknesses.

Like Waymo, the independent ADS developer Aurora (14) is extremely well schooled in meticulous design and testing of ADS sub-systems and multiple levels of integration. But they do have a greater focus on near-term use. This means defining the Aurora ODD at an early stage of the ADS design process and specific use of high definition mapping, for greater certainty in the instantaneous location of the vehicle. While the ODD is reasonably broad, there is some emphasis on the resumption of driving control from the machine when required. Frequency of disengagement of the ADS and required take-over of control have become important quality indicators for all ADSs.

Another manufacturer of even more vocational intent is TuSimple (15). They are developing an ADS for replacing the driver in over-the-road operation of heavy trucks. This ODD is described in their safety report: depot to depot haulage on pre-mapped routes, night and day, and with additional caveats of road type, topography and road laws. The ADS detects all moving and stationary objects in the immediate vicinity, predicts their future actions, and then formulates and executes a low-risk trajectory for itself. During the developmental phase, safety drivers are on board. It is intended that full commercial operation will include a fallback mode for the ADS – such as pulling over and parking – in the event of a technical failure.

General Motors have laid out the purposeful design intent of their Cruise AV (16). Like Waymo, they are pursuing the provision of on-demand autonomous rides and have carried out many miles of testing and evaluation in a challenging urban environment: downtown San Francisco. While aspects of the design process flow from the familiar avoidance of known risks, some new safety thinking is evident. The route selected by the machine may be the one best suited to the vehicle's control capabilities, rather than the fastest or shortest. The software continuously computes multiple future safe paths for the vehicle and always has a best option selected and available. If something unexpected happens, the vehicle may instantly reset its intention, even if it means going around the block and trying again. As of the time of writing, the City of San Francisco – a challenging environment for any ADS – is considering whether Cruise's operations should be placed on a shorter leash.

Mercedes-Benz worked with a group of German automakers and automotive suppliers to release a vision for AV safety in 2019 (17). They begin with the 2017 statement of the German Ministry of Transport Ethics Commission (18) that automated driving should produce a "positive risk balance" compared to human driving performance. What follows is a comprehensive step-by-step description of the designing and testing of an ADS, using an array of international standards applicable to reliable design and software validation. To a greater extent than other self-assessment reports, the German document focuses on certain levels of vehicle automation, as defined by the U.S. SAE and endorsed by NHTSA. For example, Level 3 is designed and approved in the knowledge that a human driver will be required to intervene from time to time, but this need will be minimized. These levels of automation also apply only to "our" vehicle and give no consideration to the behavior of other vehicles. This approach sheds very little light on advancing the AV safety paradigm from controlling the negative to promoting the positive.

The above-mentioned German ethics statement on AVs (18) puts forward 20 ethical rules for automated and connected vehicular traffic beginning with Rule 1: *the primary purpose of all partly and fully automated transport systems is to improve safety for all road users.* A reference to the first principles of ethics is made in Rule 2: *the protection of individuals takes precedence over all other utilitarian considerations.* Despite several strong statements about the primacy of the government in guaranteeing the safety of all AVs on public roads, some libertarian elements do appear. Rule 6 foresees an ethical issue if a fully automated system is mandated and forces the human to submit to the will of the machine. Overall these rules are tuned to well-established realities of transport safety, which are highly utilitarian, as well as providing an ethical framework for new transport technologies like AVs. For example, Rule 2 is utilitarian in the extreme: harm produced by AV driving must be less than that produced by human driving. At the current juncture, that is, impossible to prove.

It is not our intention here to figure out why AVs have not yet been commercialized at any significant scale. Nor when we will be able to park an AV in our garage or summon a ride that doesn't involve trusting the driving skills of someone we've never met. Rather, we are talking about AVs to the extent that they demand new safety thinking. They force us to be specific about what we truly mean by AV safety. Is it simply a new tool to be used in the battle against roadway injuries and deaths? Or is it a new mobility platform with much broader possibilities for eliminating or reducing the risks of driving? The ADS safety reports hew very strongly to the avoidance of failures, and risks. Some concentrate solely on "our" vehicle, while others may identify other vehicles and try to predict what they are about to do. Most are reliant on defined operational domains, and routes that are mapped with high definition. We have mentioned only one manufacturer, NVIDIA, who truly paints a picture of safe passage. It is perhaps telling that NVIDIA comes from the world of video games.

In a new world of AVs, two different notions of safety are emerging. One is that AVs will replace human driving on a large scale and will be safer than human driving. Given the high rate of crash causation currently attached to drivers, machine driving would need to be a lot safer than human driving if we are looking for a major positive impact on road safety. This is an important point – a definitive improvement in safety requires the ADS to drive almost perfectly. The other notion is that ADSs will replace humans in selected applications, provided they are safe enough; these applications may be driven by the needs of vehicle users or straight-out economics. Regardless of how much safer ADS driving is, this will not have a major impact on road safety. In either case, it will be hard to get away from our ingrained consequentialist philosophy of road safety, and the intent of official investigators to take the straightest line to the attachment of blame. Limited experience thus far shows that humans are the preferred villains, rather than driving machines or their manufacturers.

6.4 PROSPECTS FOR SAFE DRIVING BY MACHINES

6.4.1 Media Attention on AV Crashes

Collisions involving AVs tend to receive intense media attention; this leads to a significant public perception of riskiness inherent in the new technology. A prime example occurred on March 18, 2018 when Elaine Herzberg became one of the first people to be killed by an AV (19). She was crossing an Arizona multi-lane roadway with a posted speed of 45 mph, at a non-designated location. It was night-time and she was wheeling a bicycle. The protagonist was an Uber test vehicle, fitted with a developmental ADS. The ADS was engaged and a company safety driver was in the driver' seat. Among the many circumstances that were brought to light, the ADS had not recognized the vulnerable road user (VRU) and took no evasive action. The safety driver was distracted by entertainment streaming on her cell phone and over-rode the ADS far too late to avoid fatally impacting the pedestrian. The long arm of the law in Arizona ruled that Uber was not criminally responsible and charged the backup driver with negligent homicide. Uber entered into a settlement with the pedestrian's family. Uber's permission to operate their ADS-equipped Volvo SUV in Arizona and California was temporarily suspended. Following a brief resumption of testing, Uber Advanced Technology Group was sold off in December 2020.

Although this was the first high-profile fatality with an AV, the fact that the safety driver was behind the wheel diverted attention from the ADS. The developmental ADS was not ready for prime time. From Uber's point of view, the moment when human drivers would no longer be an expensive part of their business model must have suddenly receded very far in the distance. We can also see that crash responsibility as viewed by the police remained a zero-sum game, in that the investigation focused largely on the human in the vehicle – the backup driver. Some time later, Uber tried to put their actions in a better perspective of helping to develop a powerful technology that would be much safer in the long run. This single event obviously raised many questions and precipitated serious consequences that involved companies, individuals and their families, state officials, law enforcement and the legal profession.

Another tech company turned automaker, Tesla, has received much scrutiny and criticism over serious injury and fatal crashes that occur when their self-driving feature is engaged. These Tesla customers have experienced crashes when they have turned the vehicle over to the control of a simplified ADS and did not maintain sufficient attention to intervene when the ADS failed to detect unsafe situations. Tesla's efforts in self-driving should not be considered as a major part of industry's progress toward vehicle automation for the purpose of removing the role of driver error in crash causation. These Tesla vehicles were only equipped to the capability of a Level 2 AV and it should have been clear that the driver is responsible for controlling the vehicle, not the machine. But some owners pushed the boundaries of a technology that excited interest but was over-simplified and lacked a culture of safety in its marketing.

6.4.2 State Reporting of AV Incidents and Crashes

Waymo is a well-resourced and knowledgeable pioneer of machines that replace people as drivers. They have conducted much more testing of their ADS than any other manufacturer. In response to Californian regulations that require reporting of AV crashes to the Department of Motor Vehicles, Waymo have detailed a relatively large number of low-severity crashes. These are usually caused when the Level 4 Waymo vehicle obeys the letter of the traffic signal and is rear-ended by a typically assertive driver who is accustomed to cheating the red signal. Even with the tens of millions of miles covered by Waymo's ADS, serious crashes in the old world – that of human-driven vehicles – are so infrequent on a miles-traveled basis that the jury is definitely still out on the most critical question: can we prove that ADS's are going to be safer than human drivers?

6.4.3 SAFETY OF VOCATIONAL AVs

Many observers have been disappointed by the long gestation period of AVs as a commercial reality. Setting aside the long-term, mainstream prospects for AVs, how should we view the ethical responsibilities of AVs in the early, commercially driven applications? Such applications, currently in their early stage of commercial use, include ADSs for robotaxis, or rider-only (RO) AVs, and over-the-road freight haulage. In these applications, safety is not the prime reason for the introduction of vehicle automation.

In the case of RO AVs, proponents are primarily seeking reduced labor costs. Selected ADSs may be integrated in mainstream vehicles such as minivans. How much of a threat do these ADSs present for the un-automated majority? We will confine ourselves to collision threats involving this new class of vehicle and existing cars. Let us set aside the interactions of robotaxis with VRUs and heavy trucks. Previous-generation utilitarian rates of crashes and injuries cannot help us with RO AVs because there is too little data and they operate in special environments of roads and traffic. Because the driver is now right out of the frame, the responsibility for threats to other road users must be passed to either the ride service, the manufacturer of the ADS, the manufacturer of the vehicle, the insurer or the local jurisdiction (city or county) – or someone else. What is the nature of such threats? Unlike the huge asymmetry between cars and pedestrians, or cars and heavy trucks, robotaxis are basically cars, including passenger vans that meet Federal Motor Vehicle Safety Standards. They do not pose an out-size threat to other cars, as long as their ADSs drive at least as well as humans within their chosen environment.

Early in 2023, both Waymo and Cruise reported the landmark of 1 million miles driven by their RO automated driving system. In the case of Waymo, a technical paper (20) authored by Trent Victor and others was subsequently published. The paper examines all "contact events" that occurred during Waymo's RO operations in Arizona and California, using Chrysler minivans and Jaguar midsize SUVs; 20 such events occurred during the million miles of machine driving. The ODDs excluded highways and inclement weather and covered speed zones up to 45 mph. The contact events were all of a very low severity, far below typical thresholds for serious injury or death. And they were all precipitated by a lapse on the part of the other driver. The most severe event involved a Waymo vehicle slowing for a red light, and being struck from the rear by a teen driver who was looking at his cell phone. Given the design quality of the ADSs, the quality of their integration, the base safety of RO vehicles, the need for state approvals and review of contact events and crashes the prospects for further professional development of RO services are extremely good.

Waymo also operate a fleet of 48 machine-driven Class 8 freight trucks – with safety drivers – under the brand Waymo Via. Again, cost savings derived from the

exclusion of drivers' wages are the main reasons for this vocational use of AV technology. Waymo Via operations occur across several states in the Southwest, from
California to Texas. They are of tractor–trailer configuration and the tractors are
sourced from several major over-the-road brands, including Volvo and Freightliner.
NHTSA's reporting requirements for ADS driving incidents mean that several crashes
involving these vehicles have come to light. The most serious (21) occurred on May
5, 2022, on Interstate 45 in Texas when traveling in the far right lane of the highway.
An overtaking truck–trailer combination pulled into Waymo Via's lane prematurely,
contacted the left side of the cab and caused the Waymo Via to diverge from the
roadway into a wide, grassy shoulder. Tire tracks in the road shoulder show that
the vehicle continued in a straight line, departing at a shallow angle from the main
roadway, until it was redirected forward by a traffic barrier on the far side of the
shoulder area. The overtaking truck drove on without stopping.

The circumstances of this crash speak to the complexity of the chain of responsibility in automated trucking. The Waymo Via was hauling a test load, not real freight.
The person behind the wheel was an automated trucking professional, with significant experience in both truck driving and AV testing. A second person in the cab was
a software operator. Both were employed by a global transportation services company
called Transdev, known for transportation operations in many countries. The tractor
was a modified Peterbilt. The Waymo Via was traveling at a speed of 62 mph, slightly
below the speed limit. The ADS was engaged throughout the collision and the safety
driver did not resume control. The safety driver was moderately injured and taken
to hospital. The investigating police officer quickly established that the other truck
driver, who worked for a small local carrier, was responsible for the crash. Until
contacted by journalists, the officer was unaware that the Waymo truck was operating
in automated mode.

It is clear that the Waymo vehicle and its ADS behaved perfectly prior to the crash
and that it continued in a stable mode after being forced off the road. It was definitely
not responsible for the crash. It was caused by the unsafe lane-change conducted
by the other freight truck, and he had left the scene. It is noticeable that the Waymo
Via brand is displayed prominently along the side of the trailer. Could it be that a
local truck driver did not like the idea of a tech company from California wiping out
local jobs?

Even though the ADS was definitely in the clear in this case, how would the
police deal with sheeting home blame for a crash where the automated truck
was at fault? The ADS developer is clearly the instigator, but many others bear
responsibility for an 80,000 lb. truck being driven on public roads by a machine.
These include the tractor manufacturer, the vehicle modifier, the carrier or fleet
and the "third-party" employer of the safety driver and the on-board technician.
This is a big change from the traditional national approach to motor carrier safety,
administered by the Federal Motor Carrier Safety Administration (FMCSA), where
scrutiny concentrates on the truck driver and the carrier. When Waymo Via reaches
its commercial steady state, truck drivers will no longer be required. Only the carrier remains as a significant blameworthy entity. While the ADS is the replacement
for the driver and is logically front and center, it may remain far out of the reach of

investigators and law enforcement. After a century of investigating and cleaning up after road crashes, law enforcement is not looking to sheet home responsibility for complex technologies.

Once we move away from the established notions of factory installation and manufacturer's warranty, all of the players have new responsibilities. The ADS manufacturer must supply something suited to the host vehicle; an ADS originally developed for cars may need additional sensors to see well enough around a wide, tall and very long vehicle with limited field of view for the driver. The ADS needs to be integrated in the host vehicle; this may require disabling some of the existing active safety systems in the host vehicle. This integration work, and any modification of the vehicle, may be carried out by a contractor rather than the OEM. Carriers are in the habit of carefully specifying the tractors and trailers that they acquire, and tractors are often highly customized. They may have one or two preferred OEMs for reasons of fleet operation and maintenance. A given ADS developer may have preferential arrangements with certain OEMs, and this may not be transparent to the carrier. In fact, carriers will find it very difficult to guarantee the safety of an automated freight truck.

What starts out as an interesting early application for ADSs, rapidly becomes complex in both execution and responsibility for safety. And the asymmetry in collision severity between heavy trucks and personal vehicles increases the stakes for safety approvals of the ADSs driving the trucks. Prospects for the safe use of heavy truck AVs in the vocation of long-distance freight movement are therefore clouded. Even though driver fatigue is an important factor in such operations, judicious use of ADAS should be considered on pure safety grounds before resorting to the long-distance freight AV.

6.4.4 MAINSTREAM AV OFFERINGS FROM TRADITIONAL AUTOMAKERS

The above vocational uses of AV are unlikely to significantly affect road safety at the national level – either in the positive or in the negative. However, expansion of these applications in both number of sites and volume of fleets would provide cumulative evidence concerning the crash risks of AVs versus human drivers. This would be an important step in understanding the potential of AVs for pure safety. Could a motorist purchase an AV for personal use and virtually eliminate their small risk of collision with another vehicle? Not only would that purchase protect the AV owner and their family but it would also remove their slight risk to other road users.

So far, the traditional AV has usually been put forward on the grounds of convenience. It would offer to take over some of the more routine or tedious aspects of driving. Examples include stop–start traffic, control of speed and headway, and changing lanes on motorways. Pure safety is the domain of active vehicle safety rules in FMVSS – which materialize slowly – and the continued emergence of ADAS applications. ADAS has evolved into a trusted spectrum of conflict alerts, collision warnings and crash-imminent interventions. The same sensors, algorithms and actuators may be present in different combinations in all of the above vehicle systems: ADAS, FMVSS and AV.

In early 2023, Mercedes-Benz announced the availability of its Drive Pilot in the U.S. market (22,23). This option on luxury models allows the driver to be hands-off and eyes-off for periods of time, depending on road and traffic conditions, but in no case to exceed a speed of 40 mph. It has been self-certified in the State of Nevada and is expected to appear in other markets, including California. Drive Pilot exceeds the remit of other ADSs offered by automakers, which require the driver to keep their eyes on the road at all times. MBUSA's steady-minded customers may be expected to accept their cautious approach to safety and take note of Drive Pilot's rules. Drive Pilot had also been introduced in Germany a year or two earlier, with somewhat more restrictive rules of use. While manufacturers go to great lengths to make sure that these systems are safe, they are not primarily safety systems.

6.4.5 AV OFFERINGS FROM NON-TRADITIONAL MANUFACTURERS

In 2019, the Dutch Safety Board conducted a knowledgeable and practical assessment of road safety and automation in traffic (24). While their main interest was ADAS systems – which they criticized for not being designed unequivocally for safety purposes – most of the crashes in their assessment involved Teslas. It appeared that ADAS was not causing crashes of concern, but the technical blindspots and owner misuse of the separately purchased and more ambitious Tesla Autopilot did cause serious, high-speed crashes. Fundamental errors in engaging and disengaging Autopilot revealed a new source of human factors risk with AVs. And Autopilot is not designed to present an automated set of ODD limits to the driver. The Tesla Autopilot AV may be mistaken to be an automated driving feature but is in reality a driver support feature that relies upon a responsible driving person. The virtues of driving clearly mean very different things to older, more affluent, Mercedes-Benz owners and young, active owners of fully loaded Teslas.

6.4.6 AVs AND PURE SAFETY

There is no doubt that people drive more safely when their vehicles are fitted with ADAS and FMVSS systems. The widely promulgated influence of driver error on crash causation and resulting death and injury is directly countered by ADAS and FMVSS, using some of the same underlying technologies employed in AVs. This is already happening and driver errors could be better countered, more frequently, with ubiquitous presence of the technologies. And a wide swath of crash avoidance could be brought about without resorting to roadways choked with AVs. So is the widespread desire for AVs that will definitively replace error-prone drivers misplaced?

No, but it is an over-simplification. As we have said, we already have good technologies – ADAS and FMVSS – that counter driver errors. Such errors range from inattention to distraction to poor behavior. And the countering takes place when a collision is imminent. What more is to be gained by replacing the person with the machine? The answer is that only a tiny percentage of departures from good practice in driving ergonomics turn into driving errors and result in collisions. The vast majority of such departures result in conflicts that require corrective action. This is where driving machines come into their own to define AVs as the antidote to the initiation and progression of driving conflicts. ADAS and FMVSS are not able to do this.

As people drive their proximity to collisions ebbs and flows. Situation awareness fluctuates and conflicts begin to emerge. People are not good at this routine driving and are tempted to tune out because they simply lack the capacity for paying so much attention and doing so much processing for so little reason – at least no apparent reason. And none of the surrounding entourage of drivers is equipped for this task either. It takes superfast computer chips to model NVIDIA's concept of safe passage. This is where AVs are needed.

In an important distinction, ADAS deals with driver error but AV is needed for traffic conflict resolution. Does this mean that we need to make a fundamental AV choice: a person drives or the machine drives? This appears to be the implication of SAE's levels of automation (5): as we step up the levels, we cross over from people driving to machines doing the driving. But this dichotomy will probably not hold in practice – there will likely be a mix. Level 3 and Level 4 stipulate that "you are not driving when these automated driving features are engaged". In the case of Level 3 it is further stipulated that "when the feature requests you must drive"; in contrast Level 4 says that "these automated driving features will not require you to take over driving".

However, there is a little more to the current dichotomy between a person driving and a machine driving. At Level 3 it is assumed that the machine drives until it gets into difficulties and advises the person behind the wheel to take over. In a different scenario, a person could be driving but lose interest and prefer the machine to take over driving. Throughout the SAE Levels, it is assumed that automation is something that may be switched on or off. This assumption is quite fundamental. SAE is assuming that an AV is still a vehicle and has not crossed over to become a robot: the person commands the vehicle, not the machine. Control of the vehicle is shared, at least to some extent. The machine is there to support the human, not to supplant the human. Control could be traded back and forth (25) or it could be continuously arbitrated by algorithms. For example, the machine may take over lane-keeping if the driver's control is too loose; but the driver could then overcome the machine if they decided to change lanes.

6.5   ETHICS OF AUTOMATED VEHICLES

Frequent mention is made in these pages concerning road safety's utilitarian philosophy. Safety's knowledge base has been developed through study of the consequences of motor vehicle crashes. Safety's dashboard uses a variety of metrics that speak to the harm caused in crashes – including injuries and fatalities – and the circumstances and frequencies of that harm. Safety's toolbox relies on countermeasures to a large number of specific crash types. Road safety is highly pragmatic and generally avoids discussions of morality, virtue or righteousness. Because ground transportation is not a closely coupled professional system, it is difficult – and perhaps unrealistic – for road safety to adopt virtuous principles, golden rules and maxims. And the social contract under which safety regulations are handed down is not clearly articulated.

Nevertheless the introduction of AVs requires ethical acknowledgement. Virtually all forms of analysis in transportation recognize several distinct and eternal components: vehicles, drivers, roadways and the environment. Technological transformation in the 21st century has added two further components to the transportation system: Intelligent Transportation Systems (ITS), including V2X, and AVs. They demand our consideration because ITS is supporting a much greater system awareness and AV may be fundamentally changing the system. The societal boundaries of the system are also expanding. How may this impact road safety?

The distinct components of the transportation system have all been widely curated from the safety perspective; existing countermeasures tend to cluster as those relevant to (1) vehicles, (2) drivers and (3) infrastructure. The humans who happen to drive – and therefore take on the ergonomic responsibilities of driving – are clearly ascendant from an ethical perspective. In the extreme, we could view the perfect AV as a robot and invoke Isaac Asimov's Three Laws (26); this would mean that the AV could not harm a human, must take orders from humans and must protect itself. Such a spotless scenario would revolutionize road safety because each substitution of a perfect AV in place of a human driver would reduce the crash risk of that vehicle to zero according to Asimov's first law; and the second and third laws probably guarantee zero unintended consequences. This is the over-simplified vision behind the public sector's enthusiasm for transformational safety triggered by AVs.

6.5.1 ETHICS-BASED CHANGES IN THE TRANSPORTATION SYSTEM

In the real world, the addition of a cohort of AVs brings in several new interrelationships within the system; the most important is that between human drivers and driving machines when they interact in the same traffic stream – the mixed traffic situation that has elicited so much commentary in recent years. The AV needs to be able to deal with erratic behavior on the part of human drivers. Another important new relationship will be that between the infrastructure and the AV; this link is exemplified in the AV's need to respect the posted speed, stay in its lane and to stop at red lights. Such road rules are not well followed by all human drivers, so risks of collisions increase when the machine stops and the human tries to keep going. This well-known tendency has emerged from Google's, and now Waymo's, experience operating AVs in Mountain View: the AV is programmed to promptly obey the traffic signal, while the human driver definitely is not.

Taking our lead once again from the driver's ergonomic imperatives, it is essential that the AV succeeds in its transformational safety mission in all three roles of driving dynamics: responding to the demands of the road system, accommodating the actions of other road users and controlling the vehicle. These are the same responsibilities that were identified in Chapter 5 in relation to the technological paradigm of connected vehicle and infrastructure. Such AVs could be termed "ergonomic AVs" in that they relieve the driver of their more immediate dynamic system responsibilities. Certain AVs also have higher-level responsibilities relating to location, route, refueling, traffic disruptions; and the imperatives of movement that lead to a comfortable travel speed. Let us call them "mission AVs". Our discussion is limited to the ergonomic considerations of driving "this roadway at this time" because they are the driver functions that determine the vicinity of a collision, or not. In the annals of AV research and development, responding to the demands of the road system (such as they are) is relatively trivial. Similarly, control of the vehicle has been amply demonstrated. This includes pre-crash intervention at least the equivalent of a battery of well-integrated ADAS applications. However, accommodation of the actions of other vehicles has turned out to be an endless challenge. The infinite variety of situations requiring assessment and response on the part of the AV means it is extremely difficult to pre-test the safety performance of an AV. There will always be novel "corner cases" for which the ADS has neither been designed nor assessed.

From an ethical perspective, the driving demands of the road system derive their authority from the law of the land. It is against the law to speed, proceed through red lights, ignore stop signs and cross over lane lines. Enforcement may be cursory and even capricious; it may also be lacking in moral intent if it is oriented to "revenue raising". Stepping back a little, speed limit settings may be poorly conceived and traffic signals poorly maintained. Signage may have deteriorated. Funds in public agencies are short and discourage innovation. But are public agencies subject to a kind of moral law relative to roadway crashes? This is debatable. There is a big philosophical difference between causal responsibility and moral responsibility. If the traffic signal was defective, the public sector operator should bear a degree of moral responsibility because they had the institutional footprint and capability to fix it. But they did not cause the crash.

6.5.2 New Ethics of the Road System

The roles and responsibilities of road agencies derive indirectly, and very subordinately, from our social contract. And the actions of these state agencies are influenced by the economic imperatives of state industries – such as autos or tech. Is there reason to believe that AVs within the traffic stream, or whole lanes full of AVs, would change the administrative stance of road agencies? They would certainly be paying a lot more attention. Currently, some state agencies – personified by California – are setting the pace for managing AVs. They are mostly collecting information but are beginning to approve the use of AVs that come either from mainstream OEMs where "ergonomic automation" may be turned on or off, or from RO vehicles in specific missions and locations. Given current public sector belief in the long-term virtue of AVs, more and more states will be content to manage the small risks that their early-stage operations pose – and collect information.

The ethical balance will change if and when mission AVs roam widely in all classes of roadways, including arterials and highways. For sure, riders in those vehicles will be relying on the integrity of their vehicles' ADSs. But what else? They will expect the designated road network to be AV-friendly; this could start with wider and more conspicuous roadway markings but would probably include digital information broadcast from intersections and the roadsides. They will see careless driving behavior on the part of people who still wish to drive themselves as a threat; they will demand that it is stamped out, requiring enforcement of the same driving norms for both humans and machines. It seems clear that the road presence of AVs will change the transportation system for the better through their impact on road agencies' priorities and the effectiveness of traffic law enforcement.

6.5.3   New Ethics of the Traffic Stream

The massive issue of an AV's ability to accommodate the actions of other drivers – continuously managing conflicts as they develop – hinges on the rules they are programmed to follow. In the case of a robot-like AV, it will avoid contact with other vehicles in the interests of its own passengers as well as other drivers and riders – perfect. It will also be instructed by its remote controller or rider-in-chief, but nobody else; it will therefore be subject to the whims of humans and change speed or change lanes in sub-optimal ways and generate more conflict than the ideal of fully automated "safe passage" (11) – this could present problems. And in following its imperative to protect itself it may create winners and losers in its entourage of nearby vehicles; for example, it may adjust its lateral position in the lane, thereby increasing time-to-collision (TTC) for the vehicle on its left but decreasing it on the right.

Because the robot is an extreme case for the paradigm of the AV – and real AVs will allow for more human input – we can see that the mixed traffic stream is less mixed than it appears. Even the most robotic of AVs are subjugated to in-vehicle or remote human instruction. In practice, there will be exceptions to the robot laws when the robot is a commercially attractive and reasonably flexible vehicle suitable for private ownership. For example, the range or frequency of admissible human instructions may be limited. Nevertheless, AV riders potentially bear an unusual new set of responsibilities unless the "instruction channel" is blocked. All of the above brings us to a somewhat complicated view of AVs in their ability to continuously accommodate other road users when a proportion of those other vehicles are human-driven.

6.5.4   New Ethics of Crash Avoidance

We then come to the final ergonomic requirement placed upon the AV: control of the vehicle in order to lessen the imminent risk of a collision, and to avoid crashes altogether. Experience with the Electronic Stability Control system, long required under a federal FMVSS mandate, shows that vehicle-alone rapid control interventions can far exceed human capabilities. The technology is not in question, but controversy has flourished concerning the ethics of machine decisions in high-risk situations. The control actions of avoidance may conceivably involve choices when the machine has judged collision to be inevitable. There could be a machine decision to be made based on alternative recipients of harm.

Many examples have been analyzed for the specific case of AVs – for example, the work of Chris Gerdes at Stanford (27) – but Immanuel Kant's Categorical Imperative (28) provides an excellent rule of thumb. No party may be harmed in the chosen scenario of avoidance who was not involved in the original conflict. For example, when the multi-passenger AV is confronted by a heavy truck in the wrong lane, it is not acceptable for it to swerve off the road and strike a pedestrian; Kant would maintain this position even though less people will be killed by swerving.

At last we can appreciate our need for moral philosophy in road safety; not as a platitude but in order to proceed ethically with the introduction of AVs. Kant believed that each individual must be treated as an end in themselves – not a means to an end. Such hypothetical dilemmas have received a certain notoriety, but there is no sign of lab-based programmers making life choices on behalf of others. AV manufacturers are unlikely to go beyond current best practices in ADAS, and these do not extend to complicated avoidance. For example, none of the current ADAS applications involve steering control – only braking – showing that complicated scenarios of avoidance remain thought experiments, not reality. Why even raise the possibility of emergency steering – as well as braking – when it could make the crash worse?

6.5.5 CONTINUING COMMAND AND CONTROL BY PEOPLE

The insertion of AVs into our transportation system does greatly change the system, but overwhelmingly for the better. The clean safety design intent of causing zero harm to all people moving in the AV's entourage is transformational. It is more difficult to eliminate the frictional conflict within the entourage and those who serve as AV riders-in-chief or remote controllers may need training and skills. We are likely to find that road agencies have a responsibility to provide a network of designated, connected AV roadways of all classes where road rules are enforced to better align with the guaranteed compliance of AVs. Despite some expectations, the advent of AVs – replacing human drivers on a large scale – will not create a dystopian transportation system. Humans will remain in control, either by driving themselves or commanding driving machines.

6.5.6 BROADER INFLUENCES OF AN ETHICAL NATURE

The ethics of vehicular carbon emissions are already having an outsized impact on the transportation system. Vehicle powertrains and means of electricity generation have displaced road safety as a top-rank topic of public discourse. Moderation of the enormous quantum of vehicle miles traveled may be sought by governments on the grounds of carbon emissions but the reduced cost per mile of electrified vehicles (EVs) will encourage more miles driven. The increasing exposure would be a clear negative for road safety. Penetration of AVs would help to balance this situation from a safety perspective.

But an expanding cohort of AVs may tend to further increase miles traveled by increasing vehicle ownership and the number of vehicles on the road. Private ownership of mission-capable AVs may create a version of the Tragedy of the Commons (29) where the current inability of a person to drive more than one vehicle at a time goes away; such persons may see lifestyle advantage in the unconstrained ownership of multiple AVs.

Broad-based societal changes seek more options for mobility, especially in urban areas and in big cities. Individual agency and the gig economy have helped to underwrite the popularity of ride services in cities worldwide, starting with Uber and Lyft. And RO vehicle services are already one of the very first adopters of AVs. It seems likely that AVs will be viewed as an enabler of big ideas in the commons – with safety simply being assumed. The solemn duty of AVs to never harm pedestrians and cyclists may become a litmus test of their future prospects.

6.6 AVs AS A REVOLUTIONARY ROAD SAFETY PARADIGM

In setting out the U.S. Department of Transportation's desire to curate AVs as a transformational influencer of road safety, Transportation Secretary Elaine Chao said

> Today our country is on the verge of one of the most exciting and important innovations in transportation history – the development of Automated Driving Systems (ADSs), commonly referred to as automated or self-driving vehicles.... The major factor in 94 percent of all fatal crashes is human error. So ADSs have the potential to significantly reduce highway fatalities by addressing the root cause of these tragic crashes.
>
> (10)

This is pure rhetoric and AVs represent a paradigm. Paradigms are not ends in themselves – they need work. The Secretary made the connection between driver error and driving machines, but exactly how are machines going to carry out a seek-and-destroy mission on driver error?

In the 20th century the philosopher of science Thomas Kuhn (30) revealed the importance of paradigms in the advancement of science. He described paradigms as models or collections of questions and methods that a particular group of scientists exploit; that is how he believed science generally progresses. The word comes from ancient Greek and has the sense of a pattern or example that would not be a solution, but would lead an audience in a certain direction. While not being high science, road safety has utilized a number of paradigms in its history. The early days were all about occupant protection. Then there was a "paradigm shift" – a common term in the modern world – to crash avoidance. Human factors – ergonomics applied to motor vehicles – is an important paradigm, as is ITS. And connected vehicles and infrastructure, or V2X in shorthand, is a high-potential paradigm that the audience is still coming to grips with. All of these paradigms have their keepers and adherents in the form of communities of professionals. AVs are a high-profile and well-resourced paradigm unique to the technology of the 21st century.

6.6.1 TWO STAGES OF AV-LED SAFETY TRANSFORMATION

The philosophy of the AV paradigm is just as important as the technology. Considered as a new cohort of road users, the federal government wants to ensure that AVs are sufficiently safe to enter the public road system. In the absence of certainty, the final reward of system-wide safety improvement must justify the early-stage risks and the costs. Nobody is saying exactly how a sufficient fraction of the national park of motor vehicles will be converted to AVs – a big risk in itself. The federal government cannot apply its existing philosophy of motor vehicle safety regulation to AVs because standard tests are much too narrow relative to what may occur out on the road. The onus is on AV manufacturers, and some have indicated that they would take on the liability for AV crashes. Although most surveys reveal a lack of interest in purchasing AVs among automotive consumers, many auto manufacturers – but not all – have calculated a large and lucrative market. Non-traditional manufacturers and new breeds of auto suppliers see big opportunities.

The prestige of global companies, cities, states, regions and nations adds gloss to the AV cause. Some in the far-sighted digital economy see a shift in personal transportation from privately owned vehicles to on-demand ride services. All kinds of companies from retail to hospitality to communications seek to show tangible signs of virtue. In the private sector, anybody who sells or operates a bad AV may consign themselves to the wrong side of history. In the public sector, the safety allure of AVs writ large will continue to be respected. The real work continues in the vocational deployments of AVs, including RO services, ergonomic AV options in private cars and over-the-road freight haulage. In this phase, the paradigm keepers reside in the private sector. There are no short cuts. Work on transformational AV-led safety, representing a second stage, will require paradigm keepers in the public sector.

6.6.2 THE BEST AUTOMATED DRIVING MACHINES COULD DO BETTER THAN HUMANS

There is little doubt that driving machines can and will do better than humans. Waymo's automated driving system is designed to avoid contact with other road users, and it is succeeding. But our consequentialist safety mind-set is fixated on developing sufficient crash history to compare the ADS with human drivers. This tends to downplay the intense R&D effort that Waymo and others like NVIDIA, GM Cruise and Mercedes-Benz have devoted specifically to the safety of AVs. The evidence so far is that the best ADSs are anything but high-risk. The only severe crashes involve Tesla, with its doubtful ADS and foolhardy behavior by Tesla owners. To have a superbly designed, manufactured and tested ADS that never crashes into other vehicles, even though it is not always able to bypass the lapses of other vehicles with human drivers, is something to celebrate. And we appear to have several such ADSs nearing production.

The proof testing of best-in-class ADSs is in the early use cases that manufacturers have selected. They are pursuing operational cost savings through elimination of professional drivers' paychecks. This seems to work safely in some cases – such as urban robotaxis; there is insufficient evidence in other applications, such as over-the-road freight delivery. There is a big question as to who takes responsibility for ADS operations, and their level of competence in doing so. Waymo's robotaxis are proving to be safe, but what happens when others move in to provide ride services on a big scale and Waymo step back to license their superb ADS? As we see with the trucking application of the ADS, the chain of responsibility is simply too long to rely entirely on the quality of the ADS.

In order to take full advantage of the safe driving advantage provided by the best ADSs under RO usage, it needs to be very obvious who takes full responsibility. This will have to be the vehicle manufacturer. Whether they utilize their proprietary ADS, or adopt a third-party ADS, the vehicle manufacturer must be responsible for the superior driving performance of the ADS shining through in service. Any and all complications with ADS integration, such as sensor placement on different body shapes, need to be firmly in the province of the automaker. This will be a tall order for manufacturers of heavy trucks.

6.6.3 INFLUENCE OF AVs ON LEGAL LIABILITY AND COSTS

The engineering lawyer Bryant Walker Smith has devoted considerable effort to investigations of the liability of AVs and the ability of current product liability law to deal with AVs (31). On the assumption that AVs will be significantly safer than conventional vehicles, he anticipates that the annual quantum of motor vehicle crash costs will reduce; but the fraction sheeted home to auto manufacturers will increase. Product liability claims will take over from auto insurance as the main funder of collision harm and vehicle damage. In turn, this may encourage automakers to offer AV ride services rather than AV products.

Walker Smith points out that new cyber–physical systems like AVs may cause a redistribution of costs – rather than an overall reduction – perhaps negating some of the anticipated benefits. If AVs place a product liability burden on their manufacturers, and AVs cost more than their conventional counterparts, transportation system costs will increase. Some potential users may become more likely to purchase AVs than others; and a large cohort virtually never crash. It is important to remember that little is known of automotive consumer behavior relative to AVs. Would the safest drivers be attracted to AVs, leaving the riskiest drivers to do most of the human driving?

6.7 AV GUIDELINES AND TESTING

6.7.1 NHTSA GUIDELINES FOR AVs

The U.S. Department of Transportation has released several versions of its key guidance document for the design, testing and deployment of ADSs. Released in 2016, Version 2.0 (10) provides guidance for manufacturers and best practices for state legislatures. The guidance is voluntary and covers 12 priority safety design elements. It is directed to the highest-level ADSs, of Levels 3 through 5 in SAE's scheme of ADS capabilities. System safety is predicated on national and international processes respected by adjacent industries including aviation, space and military. Many standards scrutinize the design, provenance and validation of electronic control systems. This approach focuses on quality at the front end of the critical system rather than investigatory fixing of failures.

Other important safety elements include the manufacturer's nomination of the ODD and reversion to a minimal risk condition of the ADS when the ODD is exceeded. Object and Event Detection and Response (OEDR) refers to detection of, and response to, other road users and events such as emergency vehicles and road works; there is a distinction between "normal driving" – also described as ergonomic driving responsibilities in Chapter 5 – and "crash avoidance capability" where reference is made to NHTSA's catalog of pre-crash scenarios (… .)[add in ref: National Highway Traffic Safety Administration (2007) Pre-Crash Scenario Typology for Crash Avoidance Research. DOT-VNTSC-NHTSA-06-02].

Validation of these capabilities is proposed via a combination of simulation, test track and on-road testing; the possibility of independent third-party testing is raised.

The Human Machine Interface (HMI), a stalwart of human factors engineering, is envisaged within the AV as well as externally with a dispatcher or remote control authority. Further, the crashworthiness of the AV in which the ADS is installed should go beyond current FMVSS requirements to consider unusual features such as seating arrangements; consideration of the external design of the AV is also raised in relation to its aggressiveness toward other vehicles. Finally data recording is promoted for the purposes of crash reconstruction and a continuing learning environment for AV safety.

The NHTSA guidelines also invited manufacturers to submit voluntary self-assessment relative to the above safety elements. As discussed in the early part of this chapter a number of companies provided relevant documents in a variety of formats. Collectively, this information reveals some new thinking about AV safety; it is more intentional than old-line safety and less reactive to events. The documents also show that the best ADSs are tested by simulation, track testing and extensive on-road operation.

6.7.2 AV Guidelines in Other Countries

Australia is adopting an ADS safety assessment regime (32) similar to NHTSA's guidance. The elements to be considered align with NHTSA's. However, there is an important difference in that self-certification is mandatory and three levels of regulation are being set up: at market entry, in service and at the level of the states.

In 2022 the European Union (EU) implemented EU Regulation 2022/1426 (33) concerning the ADS of fully automated vehicles [no – this refers to "fully-automated" vehicles. When ratified by an EU member country the provisions of this rule become mandatory in that territory. While the essential elements of the rule are the same as those defined by NHTSA – including system safety, ODD and OEDR – it is highly prescriptive in specifying how compliance must be assessed. The scope of the rule covers fully automated dual-mode vehicles; this means vehicles that may be selected to be either manually driven or completely driven by the ADS; the scope covers vehicles carrying both passengers and freight. Distinctions are made for "hub-to-hub" vehicles on fixed routes and for automated valet parking. Considerations for features of the ODD and accommodation of other road users are more detailed than those contained in the NHTSA guidance. Requirements for the OEDR include closely delineated maneuvers where metrics of TTC must be met.

6.7.3 German Technological Development Guidelines

The automotive-minded German government saw the possibility to "significantly enhance road safety" and eventually arrive at "motor vehicles that are inherently safe, in other words will never be involved in an accident under any circumstances" (18). The Ethics Commission of the Federal Ministry of Transport opined that automated collision prevention "may be socially and ethically mandated if it can unlock existing potential for damage limitation". There may need to be limits to "dependence on technologically complex systems" and it would not be permissible to allow technology to dominate, therefore "degrading the subject to mere network elements". Manufacturers are accountable for the quality of ADSs but "the public sector is responsible for guaranteeing the safety of the automated and connected systems introduced and licensed in the public street environment". Hence a stated need for safety regulation of ADSs.

6.7.4   AV TEST FACILITIES

Automotive test tracks are commonly available all over the world and allow manufacturers to keep their vehicles away from the general public when carrying out the more extreme forms of auto testing. This may involve high speeds, rapid cornering, hard braking or sudden path deviation. Extension of this approach to AVs led to the concept of "fake cities" where the interaction of the AV with specific elements of the roadway infrastructure, traffic controls, downtown roadside furniture and VRUs could be tested. New test track concepts were needed because existing high-speed tracks do not replicate the complex urban environments that typical drivers traverse on a daily basis.

If a specific event – such as a mechatronic pedestrian walking out from between two parked cars – could be repeated many times in exactly the same way, ADS improvements would be accelerated. This and other types of controlled testing would place ADS development on firmer ground. Eventually it may be possible to have a battery of tests that would confirm the ability of an ADS to responsibly enter the public road system and begin to accumulate sufficient miles with a low rate of incidents.

Forty miles west of Detroit – surrounded by 375 automotive research centers – researchers in Ann Arbor set out to build upon the state's automotive reputation and specialize in AV safety. On July 20, 2015 the Mcity test facility was opened on the campus of the University of Michigan (U-M) (34) a one-of-a-kind test environment that includes a number of unique features designed to replicate realistic roadway, infra-structure and traffic situations. In one area is a limited-access highway, for example, while other sections resemble downtown districts and residential neighborhoods with traffic signals, mechanical pedestrians and simulated buildings.

Mcity sits on a 32-acre parcel on U-M's north campus. Not exactly a typical urban environment. But a series of unique features give the outdoor laboratory a realistic feel in a controlled environment:

- Sidewalks, pedestrian crossings, bike lanes, railroad crossings, wheelchair ramps and bus stops give streets the look and feel of urban and suburban environments;
- Access ramps, highway signage and guardrails are incorporated in a freeway section;

- Brick paver, trunk line and gravel roads provide researchers with a variety of surfaces to test automated and connected vehicles;
- Traffic signals, stop signs, traffic circles and roundabouts provide the range of intersections drivers encounter;
- Moveable building facades up to two stories high allow researchers to test how vehicle sensors and communications react to various materials and geometries;
- Mcity features a simulated tree canopy of controllable moisture content and an underpass to determine how environmental obstructions block wireless and satellite signals;
- A metal bridge surface allows researchers to test the special challenges for radar and image-processing sensors this surface poses;
- Instrumentation is embedded through the facility, including a control network to collect data about traffic activity using wireless and fiber optics, and a highly accurate real-time kinematic positioning system;
- Another distinct feature of Mcity is the presence of robotic pedestrians, which can be programmed to walk into oncoming traffic to test whether sensors can react quickly enough to automatically stop the vehicle before impact.

The AV partnerships and learnings developed at Mcity quickly led to the construction of a much larger facility at the shuttered Willow Run auto manufacturing complex in nearby Ypsilanti. This 500 acre facility called American Center for Mobility was opened in December 2017 and provides a more expansive AV test environment as well as a technology park serving a cluster of AV manufacturers. Close by, a public–private group is developing Cavnue: an eventual 39 mile stretch of AV and V2X lanes between Ypsilanti and Detroit.

A similar AV technological synergy has developed in the region of Helmond in the Netherlands. TNO is the technical arm of the Delft University of Technology and works with private partners who have expertise in testing the sub-systems of ADAS and AV and in validation of full systems being prepared for the automotive market. In addition to testing ADAS applications, including automated emergency braking (AEB) addressing pedestrians, TASS International offers a 6 km stretch of instrumented highway plus elements of urban road and intersections. Another partner, Siemens, goes deeply into AV development and validation and has patented a critical scenario methodology for AVs.

6.7.5 Testing on Public Roadways

The many problems and variations involved in testing AVs are easy to imagine. But the sheer scale of the issue is hard to comprehend. A highly credentialed project led by the German auto and aerospace industries – known as PEGASUS – set out to understand AV test philosophies that could be comprehensive and rigorously detailed at the same time. PEGASUS (35) saw a need to change the traditional mind-set of AV safety testing. They believe that the current approach in many places misses the point in that "avoiding the accidents that human drivers cause is not necessarily sufficient to reduce accident frequencies". Therefore there is no need for endless miles of testing.

According to the universal German product safety tester TUV (36), the single AV function of lane-keeping requires 117 test scenarios in seven categories to be carried out on test tracks; only 24 additional items would need to be tested on the road. The seven categories of track tests are as follows:

- Staying in the lane;
- Detection of static objects;
- Detection of moving objects;
- Following a leading vehicle;
- Entering a lane;
- Leaving a lane; and
- Field of view.

In order to rely on deterministic track tests, TUV asserts that extremely high precision and repeatability needs to be demonstrated by the ADS. For example, dynamic objects need to be identified to within 100 mm in the forward direction and 30 mm in the lateral direction. The TUV philosophy of AV testing and the disruptive vision of PEGASUS call for deterministic testing of AVs, supported by extensive use of simulation, rather than hundreds of millions of miles on the road. This requires a much larger volume of testing, at a higher level of quality, than we have seen previously in automotive safety.

6.8   IS THE AV PARADIGM SHIFTING?

After all, that is what paradigms do. The early days of AV saw great enthusiasm for the *technology* and all the known, partly known and unknown things that AVs could do for society. Not only could it transform safety, but also personal productivity and autonomy, efficient use of the road network, better use of motive energy and conflict-free traffic. Broader considerations were in the air: impacts on land use, demand for parking, city finances and employment of professional drivers.

For example, a 2015 EU–U.S. research symposium entitled Towards Road Transport Automation: Opportunities in Public–Private Collaboration (37) was predicated on expert examination of three AV scenarios: freeway platooning, the automated city center and the urban chauffeur. None of these AV operational solutions were justified on the grounds of safety, although major safety improvements were associated with all scenarios. Platooning would bring cost savings for businesses and savings in personal time. Optimized traffic flow and ease of parking were part of the automated city scenario. The urban chauffeur was seen as a boost to use of traditional transit services and shared use was an important aspect of this use case. A decade ago, the capabilities of the technology were almost a given and professional imaginations were busily engaged upon opening up new societal and economic avenues. All parts of the transportation ecosystem were asking the question: "what does the driverless car mean for us?" Nobody wanted to be left behind.

The shock of the new also generated many questions of an ethical nature. When crashes occur with AVs – even though fewer in number – would computer algorithms make secret decisions about which crash participants are harmed, and to what extent? The philosopher Philippa Foot (38) had invented so-called trolley problems to illustrate ethical dilemmas in interfering with violent sequences on behalf of others. Fears were expressed because the driving machine may have instant access to information that would speak to crash consequences for individuals. What rules would suddenly leap into action, and programmed by whom? It has since become apparent that AVs are keeping well away from such technological artistry for its own sake.

The mountainous technical challenges of AVs have channelized the research and development effort into a few vocations. This has the double effect of bringing commercial returns closer and dealing with more constrained functional scenarios. The type of automation classified as Level 3 has gained momentum, having been controversial in the early days. Sights have been lowered in the full realization that human intervention will continue to be needed in the foreseeable future. On the RO front, shared ride services lost a lot of appeal in the pandemic. And regardless of multiple economic and societal purposes for AVs, the strenuous effort to prove their quality and road-going safety has rubbed off. This may be requiring old-line safety to become more forthright and to learn from AV developers. Oddly enough, human factors principles remain as important as ever.

As we have seen with the decades-long decline in society's love for cars, people do not like to be overwhelmed by technology. We now know that drivers want more say when they interface with an ADS, not less. For the foreseeable future, AVs need to be what the Europeans call dual mode – designed for human driving as well as machine driving. People will want to be able to swap over from having to drive themselves to having the machine driving without having to stop the vehicle. This is not allowed in new European rules. The influence of human behavior in and around AVs has also increased from the perspective of AV operation in mixed traffic. Early thinking seriously entertained the concept of AV-only lanes. AV development now tends to assume that AVs need to operate safely in traffic streams and roadway grids where AVs and vehicles with human drivers use the same road space.

The ability of AVs to avoid imminent crashes now has credibility with consumers, given the success of ADAS, but it has also been made clear that the demands of ergonomic driving (or the "dynamic driving task" in some of the standards) also need to be handled completely and well by the ADS. This is much harder to define and to prove.

As with any good paradigm, there are differences in the way AV manifests itself in the United States and in Europe. The U.S. solution is starting out vocation-specific, building the knowledge base and our confidence. Europe is introducing mandatory tests that ensure extremely high quality in certain finite attributes. Another variation in the European AV solution is that AVs will most likely also be connected vehicles, becoming "connected and automated vehicles" (CAVs). While this convergence of automation and connectivity has been widely discussed in the U.S., its rationale has not been sufficiently developed. We are still at a juncture when AV and V2X are best treated as separate road safety paradigms.

It is true that some observers and commentators take the hiatus in the rollout of AV products as a very bad sign: that the promise of transformative safety will never happen. But the work continues and represents the largest R&D investment that the auto industry and its partners have ever made. Every step forward in the design of an ADS, including its methodologies of testing, represents an advancement in road safety. And there are halo effects, such as AV technologies doing double duty in ADAS applications and in traffic control systems. We are setting a high bar for the technology that we intend to rely on to drive us around; this active pre-emption means that a lot of crashes will not happen, will not need to be investigated and will not need countermeasures. For the first time, it will be possible to buy a vehicle that is, designed with a fully-fledged intent to provide safe passage.

REFERENCES

1. Othman, Kareem (2021) Public Acceptance and Perception of Autonomous Vehicles: A Comprehensive Review. *AI and Ethics*, Vol. 1 , 355–387.
2. Institution of Mechanical Engineers (2019) Public Perceptions: Driverless Cars. September 26, 2019.
3. Cunningham, Mitchell L., Regan, Michael A., Horberry, Anthony, Weeratunga, Kamal, and Vixit, Vinayak (2019) Public Opinion about Automated Vehicles in Australia: Results from a Large-Scale National Survey. *Transportation Research Part A: Policy and Practice*, Vol. 129, 1–18.
4. American Automobile Association (AAA) Foundation for Traffic Safety (2022) Public Understanding and Perception of Automated Vehicles, United States, 2018–2020. April 2022.
5. SAE International (2021) Taxonomy and Definitions for Terms Related to Driving Automation Systems for On-Road Motor Vehicles. J3016_202104.
6. Mihalascu, Dan (2023) Nevada-Approved Mercedes Drive Pilot Level 3 ADAS Limited to 40 mph. Inside EVs, January 27, 2023.
7. Nevada Regulations Chapter 482A – Autonomous Vehicles. www.leg.state.nv.us/nrs/nrs-482a.html
8. California Public Utilities Commission (2022) Statement and Map of Operational Design Domain – Driverless Deployment. Waymo CUPC Advice Letter.
9. Burns, Larry and Shulgan, Christopher (2019) *Autonomy – The Quest to Build the Driverless Car and How It Will Reshape Our World*. HarperCollins Academic.
10. National Highway Traffic Safety Administration (2016) Automated Driving Systems 2.0: A Vision for Safety. www.nhtsa.gov/technology-innovation/automated-vehicles
11. NVIDIA (2019) An Introduction to the Safety Force Field. www.nhtsa.gov/automated-driving-systems/voluntary-safety-self-assessment
12. RAND Corporation (2018) Measuring Automated Vehicle Safety. ISBN: 978-1-9774-0164-9
13. Waymo (2023) *Safety Performance of the Waymo Rider-Only Automated Driving System at One Million Miles*. Waymo LLC.
14. Aurora (2022) Safety Report. www.nhtsa.gov/automated-driving-systems/voluntary-safety-self-assessment
15. TuSimple (2019) TUSIMPLE Safety Report, Version 2.0. www.nhtsa.gov/automated-driving-systems/voluntary-safety-self-assessment. Accessed March 9, 2024.
16. General Motors (2018) 2018 Self-Driving Safety Report. www.nhtsa.gov/automated-driving-systems/voluntary-safety-self-assessment
17. Mercedes-Benz (2019) Safety First for Automated Driving. https://group.mercedes-benz.com/documents/innovation/other/safety-first-for-automated-driving.pdf

18. German Ministry of Transport and Digital Infrastructure – Ethics Commission (2017) Automated and Connected Driving. https://bmdv.bund.de/SharedDocs/EN/publicati ons/report-ethics-commission.pdf?__blob=publicationFile

19. Wakabayashi, D. (2018) Self-Driving Uber Car Kills Pedestrian in Arizona, Where Robots Roam. *The New York Times*, March 18, 2018.

20. Victor, Trent, Kusano, K., Gode, T., Chen, R., and Schall, M. (2023) *Safety Performance of Waymo Rider-Only Automated Driving System at One Million Miles*. Waymo, LLC.

21. Harris, M. (2022) Behind the Scenes of Waymo's Worst Automated Truck Crash. https://techcrunch.com/2022/07/01

22. Mercedes-Benz (2023) Introducing Drive Pilot: An Automated Driving System for the Highway. www.nhtsa.gov/automated-driving-systems/voluntary-safety-self-assessm ent. Accessed March 9, 2024.

23. Brauer, K. (2023) Mercedes-Benz Drive Pilot: The Self-Driving Car Has (Sort of) Arrived. www.forbes.com/. September 23, 2023.

24. Dutch Safety Board (2019) Who Is in Control? – Road Safety and Automation in Road Traffic. www.safetyboard/nl

25. Sarabia, J., Marcano, M., Pérez, J., Zubizarreta, A., and Diaz, S. (2023) A Review of Shared Control in Automated Vehicles: System Evaluation. *Frontiers in Control Engineering*, Vol. 3.

26. Asimov, Isaac (1950) *I, Robot*. Dobson

27. Gerdes, J.C. and Thornton, S.M. (2015) Implementable ethics for autonomous vehicles. In: Maurer, M., Gerdes, J., Lenz, B., & Winner, H. (eds.) *Autonomes Fahren*. Springer Vieweg.

28. Johnson, Robert and Cureton, Adam (2004) Kant's Moral Philosophy. *The Stanford Encyclopedia of Philosophy* (Fall 2022 Edition), Edward N. Zalta & Uri Nodelman (eds.). https://plato.stanford.edu/archives/fall2022/entries/kant-moral/.

29. Hardin, Garrett (1968) The Tragedy of the Commons. *Science, New Series*, Vol. 162, No. 3859, pp. 1243–1248.

30. Bird, Alexander (2004) Thomas Kuhn. *The Stanford Encyclopedia of Philosophy* (Spring 2022 Edition), Edward N. Zalta (ed.). https://plato.stanford.edu/archives/spr2 022/entries/thomas-kuhn/

31. Walker Smith, Bryant (2017) Automated Driving and Product Liability. *Michigan State Law Review*, Vol. 1, pp. 1–74.

32. National Transport Commission (2024) Automated Vehicle Safety Reforms. April 2024. www.ntc.gov.au/sites/default/files/assets/files/Automated%20vehicle%20saf ety%20reforms%20April%202024.pdf

33. European Union (2022) Uniform Procedures and Technical Specifications for the Type-Approval of the Automated Driving System (ADS) of Fully Automated Vehicles. Commission Regulation 2022/2144. August 2022.

34. Gardner, G. (2015) U-M Opens $10M Test City for Driverless Vehicle Research. Detroit Free Press, July 20, 2015.

35. Winner, H., Wachenfeld, W., and Junietz, P. (2017) Safety Assurance for Highly-Automated Driving – The PEGASUS Approach. TRB Annual Meeting, January 2017.

36. TUV Sud, Automated Driving Requires International Regulations. TUV White Paper. www.tuvsud.com/en/-/media/global/pdf-files/whitepaper-report-e-books/tuvsud-whi tepaper-had-regulation.pdf

37. Turnbull, K.F. (2015) Towards Road Transport Automation: Opportunities in Public-Private Collaboration. Summary of the Third E.U.-U.S. Transportation Research Symposium. Transportation Research Board, Issue Number 52.
38. Hacker-Wright, John (2018) Philippa Foot. *The Stanford Encyclopedia of Philosophy* (Winter 2021 Edition), Edward N. Zalta (ed.).

Evolution of Road Safety
From Countermeasures to Deeper Solutions

Countermeasures are the bedrock of road safety and have taught us a great deal about the causation of collisions and injuries and how they may be resisted in a professional manner. The resulting knowledge base of road safety has influenced the entire system of road transportation and has greatly affected vehicle design, the design of roadways, the behavior of drivers and the steady accumulation of technologies specific to safety. All along, the "in-house" science of human factors has developed important safety principles, particularly in relation to the vehicle.

The dashboard of road safety tells us that significant improvements in system performance have been achieved, especially in the latter years of the 20th century. But the total amount of driving has kept increasing and the result has been a steady plateau of serious injuries and fatalities – lasting for decades. That plateau is still visible among the other peaks of unintentional deaths that mar the nation. While road transportation harm is diverse in its causes and locations, this should not distract us from zeroing in on the core problem. First of all there is an unimaginably large quantum of driving in the United States. We are throwing a lot of small rocks at this, but some large rocks are called for. The transition from physical systems to cyber–physical systems is occurring throughout our economy. Our safety problem exists in our physical systems and cyber layers will make the big difference we are looking for.

Although this is far from a linear or routine task, we need not be deterred. Our painstakingly constructed knowledge of safety, along with practical ethics, instincts for collaboration and cyber helpmates are well primed for the journey. The way we go about designing the cyber–physical system should be based on the principles of human factors engineering (HFE). These principles have gone missing in recent years, having previously made a decisive difference in-vehicle design. Safety's prime object of fatalities in motor vehicle crashes – a compelling motivation in itself – is really the tip of the iceberg in our attempts to turn the tables on collision causation. Technological paradigms have the required breadth and depth to transform the quality of driving to a national status quo of *safe passage*. And the powerful influence of automated vehicles (AVs), alongside connected vehicles and infrastructure, has been revealed as quixotic.

DOI: 10.1201/9781003483861-7

7.1 BIGGER SOLUTIONS WITH MORE RAPID APPLICATION

The massive anonymity of driving in the United States is hard to comprehend. Annual miles of road travel are several times that of air travel, for example. And we have no information about who travels or where they travel. But we know that virtually everyone does it. Despite the many influences tending to resist endless driving on the nation's roads, it continues to increase. Driving is mostly done in support of Americans' daily channels of living and working but is not a coherent activity in itself. It is not necessarily planned for but there is a generation-spanning sense that it will continue throughout one's entire life.

For the majority of the population driving is regarded as perfectly safe and *is* extremely safe from an individual perspective. But the risk of involvement in a collision and then being injured or killed is not zero. With such an extreme quantum of driving, fatalities are produced by the ground transportation system in a large number of separate events that occur all over the country. These events are small, violent and may be unheralded but occur on such a steady drumbeat that road fatalities accumulate to be the equal of any other major cause of unintentional harm. And fatalities are only a small and most tragic part of the story.

7.1.1 All-of-Crashes Approach

The events that cause these fatalities are part of a much larger spectrum of collisions that damage vehicles and roadways and cause non-fatal injuries with a wide range of severities. Injuries range from the inconvenient to the chronic, debilitating and livelihood-disrupting and are costly. The people causing and experiencing all of these crash events each year are a small percentage of the population of drivers. The driving population holds the very thin qualification of a driving license, learns by experience and shows extreme variability in skills; but drivers are only considered risky in their teen years and when they become old. The total cost of the huge number of damaging events is comparable to the all-up national expenditure on road repairs or total losses in storms, high winds and fires.

For the year 2022 National Highway Traffic Safety Administration (NHTSA) reported over 42,000 fatalities and over 2 million injuries in road vehicle crashes (1). Extending the picture to property-damage-only crashes, NHTSA data for 2020 (2) shows more than 3.5 million PDO crashes amid a grand total of over 5 million police-reported crashes. That total includes crashes causing fatality, injury and damage. While it is true that we have more data about fatal crashes, they are a slippery target because they represent a small proportion – less than 1% – of motor vehicle crashes; and we tend to be very much in the dark as to what singles them out in terms of severity. If we wish to eradicate fatal crashes, the only sure way is to wipe out all 5-million-plus crashes. Once we know that a crash is indeed turning out to be a fatality, it is much too late to stop it.

7.1.2   Changing the Dynamic Margin in the System

At that point in our thinking, the elimination of fatalities becomes less a moral crusade and more a necessary change in the transportation system. We know that roughly the same amount of harm will occur next year as occurred this year – that is, what well-oiled systems tend to do. The only way to change road safety is to change the system. The seeds from which road safety grew – the study of the worst vehicles in the most serious crashes – have lowered our sights so much that we only consider one vehicle per event: our vehicle, the protagonist or the host. But millions of violent events happen when vehicles crash into each other; and this occurs when they lose their margin of safety, vehicle-to-vehicle. Unfortunately the limitations of crash data based on the evidence left scattered on the roadway have so far kept causation – especially multi-vehicle causation – hidden from our view. Using current data, the unresolved conflict would be a one-sided and static narrative at best.

Each and every vehicle must be able to sense and maintain its margin of safety relative to other vehicles. It needs to do this in a manner that is acceptable to its driver. Such a feat of human engineering is more than technology and requires a working knowledge of human strengths and weaknesses in the dynamic driving task (DDT). The current technologies of Intelligent Transportation Systems (ITS) are relevant to this difficult mission for the new age of road safety: namely, human factors engineering (HFE) calling forth mastery of dynamic margin in common roadway types and driving environments.

7.1.3   Safety at the Source

Motor vehicle safety regulation provides a big model for the enterprise of road safety. With a single stroke at the point of sale, the motor vehicle is given a quantum safety improvement; and that improvement endures for the life of the vehicle. Even when someone else buys the vehicle, the benefit continues. But the fact remains that only a relatively small percentage of the vehicle fleet turns over each year. So it does take decades for that quantum of safety to penetrate the transportation system. And it also takes years for a new regulation to be developed, negotiated and come into force. In the current era, such regulations are aimed at crash avoidance and created in software. Because they take control of the vehicle from the driver in an emergency it takes time for the parties to be sure that they will always be a force for good. For these reasons, safety regulation probably lacks the reach and agency required to improve the dynamic margin of the transportation system.

7.1.4   Harnessing of ADAS

It is said that justice delayed is justice denied. In the case of safety technologies, delays are only part of the problem in putting good solutions to work. The regulations are fastidious in avoiding prescription. The tradition is that regulations specify a certain level of performance but do not require any particular methods – and especially do not mention specific products. The emergence of advanced driver assistance systems (ADAS) has proven that such technologies are effective, well-liked and may be developed quickly. There is already a long list of such applications, each representing a specific emergency situation. But the automakers act as gatekeepers in that they market ADAS to their customers much as they would market any other optional extra. People do not perceive ADAS as a major threshold in vehicle safety and may not appreciate the real safety value. Currently there is insufficient aggregation and data to put ADAS forward as a big solution. But there is potential for rapid expansion in the applications – meaning coverage of more crash types – and a lot more clarity about their value.

7.1.5   Lessons of AVs as Work in Progress

The new technologies of AVs are ripe for exploitation as individual solutions, even though the market entry of AVs as entire products is not yet with us. The question of improvement penetrating the entirety of driving has changed. It is not just a matter of the rate of turnover of the vehicle fleet limiting the overall availability of crash avoidance. The target needs to be enlarged. Traffic conflicts happen a lot more often than pre-crash emergencies. By setting the solution threshold at a lower level of activation many more paths to an eventual serious crash may be blocked.

There are other lessons provided by the hard-won progress so far with AVs. Instead of the slow-burning regulatory approaches of the past, AV development was stimulated by a set of *challenges* that relied on reputational rather than monetary rewards. The underlying stimulus was excessive losses of military drivers in theaters of war. There was an urgent need to move strategic materials along roadways contrived to be lethal. The big idea of the autonomous military transporter was then pressed into service in the much more complex and sensitive environment of public roads; a market opportunity well above and beyond military applications. These clever machines are not commodities and may never be commodities. Currently, it is unreasonable to expect rules to set the performance of an AV; that may be what one usually does with mass market products of societal significance. But it won't work with AVs. For the AV, its provenance and the merit of its creators are the main factors in its success. Much like the example set by people of virtue; examples of excellent automated driving systems (ADSs) have emerged and should serve as exemplars to be emulated. That is unlikely to happen under an exhaustively even-handed approach by safety agencies that still want a dominant national role.

Perhaps the greatest safety contribution of AVs at this juncture is conceptual. AVs are not able to act solely in a reactive way – avoiding certain things – but need to be creative in providing movement. The idea of "safe passage", which appeared courtesy of the chipmaker and AV developer NVIDIA, speaks to the safety cocoon or safety margin that the ADS seeks to maintain relative to other vehicles. Other notions have been introduced that acknowledge the various machinations of ADSs in continuously arbitrating their cocoon relative to other vehicles. Similarly, researchers have articulated some of the ethical decisions that ADSs must make as they contend for road space. Bringing these ideas out into the open makes us realize that we are not able to fully express how human drivers go about safe passage, even though they must exert themselves extensively for this purpose.

7.1.6 ENDORSEMENT OF GOOD TECHNOLOGY

Government actions in road safety have a tradition of studiously avoiding endorsement of specific technologies and products. Through their reliance on standards and regulations, safety agencies refuse to provide opinions. In order to re-human-engineer the inner workings of the transportation system, good technology needs to be made available in smaller packages and more frequently. This requires a lot more than the provision of frameworks and general guidance. Everyone needs to know that safety is not going to change significantly until all of the good pieces are brought into the light to create a better system.

7.1.7 SYSTEM IMPROVEMENT

The philosophy of Vision Zero introduces the concept of system responsibility for the causation of motor vehicle crashes. This comes about through a desire to shift the current emphasis on driver error to a more inclusive view of causation. In that view, the roadway should be held partly accountable because it could be designed to be more forgiving when a driver does make an error. But in moving from sectoral causal factors limited to drivers, vehicles or roads to a system responsibility, it is not at all clear that the interactive behavior of the system becomes the new focus. Despite the new thinking forced upon us by AVs, we are probably not ready for a fully interactive taxonomy of driving.

The newly important idea of the DDT is a system interaction in close proximity to the critical sequence of events: namely, conflict between vehicles moving to the state of pre-collision, emergency actions and the crash event. At the other, more traditional, extreme the behavior of the traffic stream is treated wholistically and all we need to know is the speed of flow and the density of the traffic. When it comes to intersecting traffic streams, traffic controls are supposed to prevent conflicts between vehicles. We therefore have a limited and discontinuous impression of the driving system. That is why AV developers have been forced to develop a complete model of the DDT.

What levers of improvement do we have? Speed limits immediately spring to mind, but they are more important for reducing the severity of injuries than for avoiding collisions. Traffic signal improvements, including adaptation and coordination, are one option. And of course more coverage of roadway intersections would be an important factor. When we come to roadway factors, we know that roadway class affects safety but the infrastructure does not currently hold up too many avenues to system improvement.

Safety has arrived at the point where exciting transportation system improvements present themselves and will change road safety for the better. And there are not a lot of current options. The first change is traditional with a new twist and falls mainly on those responsible for the national roadway network. Broader signal coverage is needed of the nation's trafficked intersections, together with ITS that increases awareness of the signals. Only a small minority of intersections currently has signals and signals are not sufficiently effective. The second change is all new: preservation of the dynamic margin between vehicles in the traffic stream. This represents the biggest human factors challenge of the technological era of safety.

7.2 THE ENDURING VALUE OF HUMAN FACTORS APPROACHES

HFE, or ergonomics, has long been the lifeblood of road safety. The development of HFE coincided with the establishment of the road safety profession in the second half of the 20th century. How exciting this was – the idea that human performance could be optimized. And that the relationship between the human and the tool – and eventually the technology and the machine – had somewhat fixed attributes that could be worked on and improved. Most importantly, the machine could be designed to suit the capabilities of the humans that used it. This attractive principle would bring lasting benefits through the entire life of the machine. Looking at it a different way, a machine could help a human carry out a task that they could not do on their own. The person–machine relationships coming under the heading of ergonomics came to include the physical and the cognitive, and eventually the intelligent.

7.2.1 HUMAN FACTORS NOT FULLY UTILIZED IN SAFETY

It is obvious that HFE made a huge contribution to the safety of the vehicle over many decades. But its relevance to the safety of the highway system and road environment has only been recognized recently. In particular, the Dutch approach to sustainable safety – where roadway elements signal their design intent to motorists – was only articulated in 2009 (3). According to Jan Theeuwes and his colleagues, "design principles can reduce the probability of errors while driving". But the unrecognized need to develop golden rules for human factors in road design is only a first step. What of the behavior of traffic? Is it possible to design traffic? It would be unthinkable to attempt such design outside the realm of HFE. And yet Theeuwes goes on to point out that "traffic engineers primarily solve problems without consulting experts in human factors".

As we shall see, human factors has often been neglected throughout road safety and even within the home field of road transportation. Across the many functions of the road safety enterprise, there are numerous examples that remain untouched by human factors: speed limits, traffic signals, traffic rules, driver licensing, ITS, crash databases, on-road enforcement, liability and insurance, safety campaigns, education, national road safety strategies and new urban mobility modes. Virtually all of these things are aimed at the humans who drive motor vehicles, but none of them give active consideration to the talents and limitations of humans who have the difficult task of safe driving. This may have something to do with the fact that many of the functions have an adversary flavor, or could become adversarial, depending upon the circumstances. Better to avoid the subject. Or maybe we do not want to be reminded just how difficult many of these functions really are; for example, detecting a pedestrian in an unexpected location on a wet night.

One interesting example against the trend is the case of the hours-of-driving rules for long-distance truck drivers. These rules are subject to extensive research on the effects of driver fatigue on performance and driver needs for rest breaks. They are heavily contested and controversial within the trucking industry. But these rules have a critical impact on drivers' livelihoods and carriers' businesses. Is that what it takes to bring in the big guns of ergonomics? Human factors has some ethical overtones – "this machine makes the operator's life safer" – but the moral worth of ergonomics has been little explored. This is probably a major oversight when the subject at hand is the heavyweight global issue of motor vehicle safety.

7.2.2 APPLICATION TO DYNAMIC DRIVING TASK

Even if we restrict our consideration to the DDT itself, human factors still looks very much under-appreciated. In Chapter 5 we described the driver's ergonomic responsibilities as: responding to the signals and signs of the roadway, accommodating other drivers' actions and controlling the vehicle. Only the latter – braking, turning and avoidance – has been thoroughly aided by human factors approaches. But even then, the technologies of ADAS imply a critical view of the driver – that they are too distracted to take proper care of emerging conflicts. The full application of human factors principles to assist the driver – not to underestimate or blame them – is needed through functions including traffic control and ITS.

Another view of this failure of safety to be sufficiently rational is provided by distinctions in the harm-producing events. Early safety started with addressing the "second collision" – the human being injured by the interior of the vehicle. The science required here was more biomechanics than human factors. Then safety's attention turned increasingly to the "first collision" and how it could be avoided or lessened in severity. Whether the technology warns the driver or jumps in with avoidance, it minimizes its interaction with the driver and affords them little credit. But there is no doubt that the science of human factors contributed to avoidance of the first collision, even though it could have done more.

Now we want to avoid the traffic stream conflict that precedes the first collision. Drivers have no chance of monitoring all of the time-to-collision vectors between their vehicle and the other vehicles in their entourage or intersection. This requires HFE all the way, in order to ensure human-friendly traffic. Some of the technology may resemble the AV but the philosophy is completely different. The intent is to reinforce the agency of the driver, not to bypass it.

7.2.3 LIMITED IMPORTANCE OF RAW VEHICLE SPEED

Scrutiny of national safety strategies around the world reveals a high dependence on reducing vehicle speeds – as a major tool of road safety. This is often planned through reductions in posted limits and automated enforcement. But the sequence of crash events is not uniformly sensitive to speed. The first two events – the unrequited traffic conflict and the "first collision" – are less sensitive to speed. It is only the third event – the "second collision" that is highly sensitive to vehicle speeds. This is the main reason for the prominence of speed in road safety campaigns. It requires excellent crashworthiness designed into the vehicle and is not very dependent on human behavior now that drivers and occupants have developed the good agency of seat belt wearing.

Turning now to our three ergonomic driving responsibilities that make up the DDT, do they have blanket dependence on vehicle speed? Let us consider each in turn:

1. The response of the driver to road signs and signals is not highly speed-dependent, although the need to stay in the lane may become more critical at high speed.
2. <u>Accommodation of other drivers</u> requires harmonization of speeds within the entourage, but not lower speed travel per se. That is, the average speed of the vehicles in the entourage is less important than the variability of individual speeds. This is where the accommodation between vehicles becomes critical and may exceed the ergonomic capabilities of the human driver.
3. <u>Control of the vehicle</u> is much improved at a lower speed. The handling and braking of the vehicle at high speed requires greater skill in-vehicle control, but HFE on the part of vehicle manufacturers has done much to complement and extend the limitations of drivers.

Considered in this way, a complete human factors accounting of the influence of speed on the DDT reveals an uneven portfolio with a few red flag functions, but a number that are more mixed or even moot. It therefore seems unlikely that artificially low speeds – brought about through extraordinary coercion on the commons – will bring us to a condition of zero harm.

Our current standing on the trajectory of road safety improvements brings safe passage – the necessary accommodations between vehicles traveling in a similar but unharmonized manner – into sharp relief. This is not really about speed as such, but about contention and adjustment. Contention speaks to pressure exerted by one driver to improve their position in the traffic stream, perhaps to maintain their margin of safety or to gain separation from a perceived bad actor. Adjustment means the granting of space to a contending vehicle; we have seen that the ethics of automated driving favor space being given, not taken.

7.2.4 IMPORTANCE OF RELATIVE SPEED AND CONTENTION

The DDT requires a sophisticated perception of space and speed; that is, the sensing of distance and its first derivative by time. Within a given lane, the only way that space can be granted is by a subtle change in speed: slowing down to provide more distance to the lead vehicle (much more likely) or speeding up for the following vehicle. If these other vehicles are also adjusting their speed, our protagonist needs to adjust the *relative speed* which may be different for different target vehicles. If we add the question of changing lanes, contention is virtually assured and the degree of difficulty is multiplied. It is now more difficult to judge available space and the intentions of other drivers. ADAS applications such as side object detection have become popular to assist in a difficult task that could become stressful. However, ADAS may be more about "what can we sense?" than "what would help the ergonomics?".

7.2.5 THIRST FOR SITUATION AWARENESS

The ultimate expression of V2X, where every vehicle is equipped, would provide our protagonist with a moving heat map of contention and model adjustments. But there is still an unmet need for human factors principles to utilize this information in assisting the driver with the very things they have the greatest difficulty in doing. There is no shame in this – it is completely understandable. But for too long we have pretended that it is normal for relatively untrained people to glide along and quite regularly hope for the best. If their unflappable demeanor is an act, does that act extend to tuning out and seeking distractions? Would the restoration of authentic agency to the driver, at the same time experiencing the respect of human factors design, reduce the "driving errors" that beset road transportation?

Situation awareness in driving is currently a mirage. Driving situations are often so complex and fleeting that, after a 30-minute drive, people are said to have no memory of the trip. People cannot process the situations piling one on top of the last, let alone retain or articulate them. Although we do not know for sure, the many insights that have accumulated through the history of road safety research imply situations on a continuum of contention. That escalating spectrum would include cooperation, accommodation, contention, intrusion, conflict, emergency, avoidance, collision and injury. Road crashes do not appear from nowhere; there is a sequence including a poor choice that narrowed the options and increased inevitability. The application of ITS, including V2X, could revolutionize situation awareness if it occurs within a human factors framework designed with good driving agency in mind.

7.3   THE EMERGENCE OF TECHNOLOGICAL PARADIGMS

A decade ago, the U.S. transportation research fraternity was grappling to come to terms with the new (4). This may sound like a contradiction. How could those on the cutting edge find themselves almost blind-sided by a spate of new technologies? The mechanism of acquiring new knowledge through posing relevant research questions has served us well and continues to do so. But our method had fallen behind the reality of unutilized technical capabilities that had emerged very quickly. These potential bearers of innovation had appeared suddenly, not gradually. Not the least of it is that they are mostly cyber–physical in nature; that is, digital systems were transforming existing physical systems. Such means of rapid and affordable innovation are hard to ignore. For example, a *smart city* is still a city with new layers added. Transportation is an attractive arena for early adoption of cyber–physical technologies because it is data-rich; it is also fertile ground for people-oriented narratives that can be turned into compelling solutions.

There is a name for this process that goes back to the ancient Greek: what was appearing in several aspects of transportation resembles a *paradigm* in that it is a model or toolbox plus an audience. Once set in motion by qualified champions, the paradigm could manifest itself in different solutions. Those solutions bring together multiple partners, points of view and capabilities. A decade ago, the U.S. Transportation Research Board (TRB) used the term "hot topics" in order to identify promising avenues for deeper engagement (4). These topics were paradigms in various stages of recognition and development. As expressed by the TRB in 2016: "The topics are all technological, but include elements of service provision, business models, and broader impacts, including community value and societal impacts. Even when we speak of technologies, we mean systems, often with multiple layers of base technologies". Foremost among these models of transportation advancement was *connected and automated vehicles*. These are two separate topics but researchers were quick to recognize the synergy between them. Both paradigms have produced road safety solutions that were not necessarily expected and have surprised with their depth of influence.

7.3.1 CONNECTED VEHICLES AND INFRASTRUCTURE

Vehicle-to-Everything communication (V2X) is a direct descendent of ITS and has suffered from the same problem of being all things to all people. As proposed in 2005, it was about safety, but also mobility. It would be initially be championed by the infrastructure operators, but the vehicle industry may be willing to take over; more recently, should it be part of the infrastructure after all? Even then, the secret sauce would be provided by the auto industry's backroom for crash avoidance technology. Contradictions abound. V2X has remained a safety paradigm without the strong adherence of road safety professionals nor human factors professionals. Its terminology is convoluted and is not clear even to experts. This is not helping us to get beyond the notion that automotive consumers do not like "gadgets" – that they are annoyed by little technologies that may create false alarms.

Thanks to the persuasive influence of the V2X paradigm to mobilize large field operational tests (FOTs) good ideas materialized. The idea of cars talking to each other and to the infrastructure, and of course including drivers in that loop, became the focal point for practical revelations about safety. Two distinct safety solutions may be emerging. But they are not what was expected. Even though V2X safety started out as an affordable way to spread crash avoidance applications throughout the driving population, it has its drawbacks in this context. Not the least of these is an inability to integrate the application with the vehicle controls. For example, in situations where emergency braking is likely to be needed, V2X warnings are going to be insufficient. And vehicle-to-vehicle (V2V) positional accuracy may not be sufficient for near-crash applications. But alternative perspectives have arisen through FOTs; when a significant number of drivers are living with the *paradigm*, rather than the first-up solutions, many imaginations are lit up.

Field operational testing has shown that solutions to the driver's absence of situation awareness do exist; but that *absence* first needs to be recognized as a major impediment to road safety. Even with the best wireless technology operating in a pristine licensed band, the dynamic relativity of vehicles on a multi-lane roadway is a wicked problem. If the current technology only gets it right half the time, what chance the unassisted human behind the wheel? Severe driver quandaries also arise at intersections with turning vehicles; FOTs have already come close to a solution with the Intersection Movement Assist use case.

FOTs show that drivers do care about the way they drive, but they have very little sense of driving criticality: what do I need to be aware of, and how well am I doing? It may be a stretch to conclude that the sensory intimidation of unassisted driving leads to tuning out, but the extent of inattention and poor driving behavior in crash causation cannot be ignored. Much of this safety solution calls for V2V technology. It operates in the dynamic zone of contention and requires high authenticity and accuracy. Less relevant is the split-second immediacy of pre-crash and emergency. A different kind of solution is proffered by vehicle-to-infrastructure (V2I) technology. The most important V2I use case arising from V2X FOTs speaks to the impending red light at a signalized intersection and lets the driver know that they do not have time to proceed through the intersection – and that they should stop.

Simplicity is important and safety use cases should adhere to quality rather than quantity. Otherwise the driver could be confused by multiplicity – either in different functionality in use cases or as to which among multiple "other vehicles" could be coming into conflict. The very idea of objectivity in traffic conflict has appeal. Drivers appreciate clarity; for example, being told which vehicle they should be wary of. And FOTs that field a limited number of V2X-equipped vehicles have noticed that unequipped vehicles are affected for the better. Drivers seem to like to be respected through the provision of quality "inside" information including traffic signal phases. With V2I there is an opportunity for the operators of the infrastructure to reach out directly to a cohort of customers who normally remain behind the curtain.

7.3.2 Automated Vehicles (AVs)

AVs represent a paradigm of liberation. Whether we seek to avoid being driven by another fallible human or wish to multi-task while traveling in our personal vehicle, automation frees our time and attention – and does it safely. We are attracted to the AV paradigm on the basis of freedom, more so than safety. But safety is simply expected in this technological era wherein so many desires are fulfilled. The question is not so much whether AVs always drive safely – although that is an important question – but is more to do with what AVs can do to back-fill our continuing quest for humans to drive safely. In figuring out how AVs *must* drive we also learn how humans *should* drive.

The many experiments needed to tease out the meaning of the good driving that is instilled in AVs also tell us what we, who have been doing all of the driving, should have been doing all along. In the process of developing their ADSs, manufacturers have needed to invent the functions of driving. This has included quasi-human concepts such as the "ego" vehicle, "claimed" road space and the "safety force field". The whole idea of *safe passage* provides a vital model for a new era of safe driving by intent. And this may not require a wholesale swing to full-machine AVs. With the help of V2X humans may be able to mimic the driving functions invented for AVs and avoid many errors in the process.

Such is not necessarily the gift that AVs were expected to bring to the table of road transportation. AVs were supposed to simply replace error-prone human driving on a large scale, and thereby transform road safety. As a solution inspired by the paradigm of AVs, this approach remains incomplete. But the main enemy of road safety, driving error, is a human factors problem; there has been little attempt to engineer our way out of it. Before we discard the human driver, we should perhaps find ways to provide them with clearer situation awareness upon which to perform the act of driving more reliably. And from a practical point of view, more than 200 million American drivers will not be easy to bypass, even if that turns out to be the right course of action. It is likely that habitual human drivers will accept technological assistance if it is useful but will not just flip the switch to automation. Such a transformation would need to occur in steps, perhaps including certain stages of shared control. For example, the machine could simply protect the claimed road space and leave the rest of the driving to the person behind the wheel.

Meanwhile, vocational AVs that save time and money have obvious relevance to road safety. But they still need to prove that they are safer than humans in their chosen environments; once this happens it will represent a big step forward. At that point they may transform industry sectors but they will not automatically transform road safety.

7.3.3 THE POWER OF THE PARADIGMS

The 21st-century paradigms of V2X and AV are needed to transform the quality of driving. The quantum of driving is continuing its upward trajectory and most people wish to continue as drivers-in-chief. Road safety has little option but to find ways of eliminating driver error. This requires deep application of the science of human factors in the development of solutions based on these important paradigms. The tendency for road safety to break into the separate constituencies of driver, vehicle and infrastructure led to under-utilization of human-facing methodologies; and our thinking became too narrow. In the parsing of causal blame for collisions between these sectors we lost sight of the pre-eminence of the human and our duty to support the human before all else.

The ancient philosophical concept of *techne* – an antecedent of the word technology – speaks to the application of practical knowledge as an art or skill (5). We should not be too quick to dismiss the agency of driving without considering it as an ingrained part of the human psyche. Because it supports so many of the most important strands of our existence, it would be surprising if we took no pleasure or pride in driving. It is not something to be bypassed. Plato (5) discusses *techne* in a wide variety of crafts, including medicine, piloting a ship and chariot-driving. Each craft has a human practitioner and a goal; and that goal is practical and tangible.

In this frame, the *techne* of driving a car obviously entails a driver and movement; and there are practical reasons for the movement – it is not simply discretionary. Plato also says that the practitioner is able to give an account that informs and guides the skilled practice. This raises an interesting question about the extent to which the modern car driver is a skilled practitioner and is able to offer an account; for example, one that would help other drivers. But Plato stipulates that a practitioner must be free in order to provide an account. To what extent is a driver able to access and express free will while driving on modern roads among modern traffic? This is not something we ponder very often.

The futurist author Isaac Asimov (6) provided the Three Laws of robots; these are often quoted in discussions of the ethics of AVs and are referred to in Chapter 6. If we agree that the *techne* of driving is important perhaps there are Three Laws of Drivers waiting to be expressed. After Asimov, here are his robot laws followed by my versions relevant to human driving. It is assumed that the robot laws would apply to any AVs in the traffic mix.

Three Laws of Robots (Asimov):

1. "A robot may not injure a human being or, through inaction, allow a human being to come to harm.
2. A robot must obey the orders given it by human beings except where such orders would conflict with the First Law.
3. A robot must protect its own existence as long as such protection does not conflict with the First or Second Law".

Three Laws of Driver Practitioners:

1. A driver may not injure a human being or, through inaction, allow a human being to come to harm.
2. A driver must obey the laws and instructions given it by the regulated entity of infrastructure and traffic except where such laws and instructions would conflict with the First Law.
3. A driver must provide an account of their freely chosen practice of driving to at least one other practitioner of driving, including their commitment to:
   a. Accommodate the actions of other vehicles
   b. Cooperate with driving machines.

In this thought experiment, Laws 1 and 2 seem fairly obvious but Law 3 much less so. The idea here is that a practitioner of driving must be able to give an "account", otherwise they could be a machine. That account should be absolutely their own and instructive, in some chosen fashion, to other practitioners. The reference to driving machines in 3(b) would cover other vehicles being driven by ADSs as well as ADS-like devices in their own vehicle. Clearly many questions arise but some form of articulation along these lines will be needed when ADSs arrive in our vehicles. It is interesting to note that the fictitious date Asimov applied to the promulgation of his Laws was 2058 A.D.

The unleashed paradigms of V2X and AV have involved significant numbers of people in personal FOTs and in vocational trials. It has become clear that foisting un-human-engineered solutions upon drivers does not work. The use cases and applications that make sense to a technical committee need to also pay their respects to the drivers-in-chief. Iteration is essential and the automakers' mantra of multiple learning cycles with real users should be adopted. V2X and AV both have broad constituencies of adherents led by a small number of champions. Those champions may change along the way, depending on the solutions that are emerging. The group must have strong government engagement and support. Much remains to be done to exploit these methodologies.

V2X and AV are the paradigms that speak most directly to the DDT and therefore most squarely address safety. They are products of cyber–physical fusion. V2I originated through the overlaying of sheltered wireless connectivity across the road network. V2V appeared through the placement of wireless – hopefully the same wireless – in vehicles. AV required the creation of its own cyber–physical system in the form of the ADS. The driver had become an ad hoc cyber–human system through various assistance technologies such as ADAS. In an interesting twist, aspects of AV technology may play an important role in this system before it is ready to create its own cyber–physical system. The paradigm of AV has therefore split. In addition to the paradigm of the stand-alone AV, the paradigm of the shared-control AV may well develop its own constituency.

Is it possible that the layered cyber–physical system is just too convenient? In the context of AVs it has been suggested (7) that such systems may simply shift problems around and not really solve them. Or that externalities attached to the legacy physical system will not be left behind but will be carried over. In a system containing stand-alone AVs, there would be an unwanted impact on congestion and travel time with overly fastidious AVs; there may also be more vehicles on the road at any one time. And there may be less willingness to pay for public infrastructure among vehicle owners. Shared-control AVs would be a hit with the vast majority of drivers who may be willing to pay a little more for much-improved driving agency. V2I-enabled infrastructure may cause a redistribution of traffic patterns toward the equipped roadways; this would be a positive for safety as well as emissions from traffic. V2I will also be greatly appreciated by the drivers-in-chief who would be pleased to see 21st-century upgrades on their behalf. V2V-enabled vehicles would of course be attracted to the V2I roadways and intersections. Enabled vehicles may also tend toward longer trips and long-distance routes.

None of these potential side-effects appears to overwhelm our reasons for pursuing a thorough overhaul of the nation's quality of driving, through large-scale technological solutions. The main issues are attached to the widely proposed but still distant solution of the stand-alone AV. The fastest and biggest change will be the invention of *national safe passage*; this requires the planting of V2I in the infrastructure in order to serve the agency of driving.

7.4 THE RIGHTS OF PASSAGE

The core of road safety is the driving of motor vehicles upon the network of public roads. First and foremost, road safety rises and falls on the quantity and quality of that driving. There are many other things with which road safety concerns itself but its sense of self-worth hinges on these two inescapable realities. The quantity of driving continues to increase and there are few prospects for reducing it, even if we wanted to. Driving is not discretionary, it is necessary based on the lives that hundreds of millions of people wish to lead. Therefore the only way to transform safety is to improve the quality of driving. Much has already been done through the 60 or 70 years of the age of road safety; but now the curve has flattened along with many people's expectations.

In an age where technology is invited into our homes, cars and wallets, not to mention many forms of public infrastructure and services, people expect new possibilities. Cyber–physical systems abound where cyber layers overcome the limitations of familiar physical objects; we have seen trade-offs and unintended consequences fading from view. Can people have their cake and eat it too? In the case of ground transportation, new mobility options are tailored so closely to wants and needs that they enjoy instant acceptance; they may be loved. And they are simply expected to be safe. It is therefore becoming an anachronism to battle with the residual harm caused in motor vehicle crashes. This must be a hangover from the 20th century.

Road safety is reactive by nature because it was birthed by the medical profession that is charged with picking up the pieces in the often-devastating aftermath of motor vehicle collisions. It is now time to fully unpack the 21st-century paradigms of transportation based on cyber–physical systems. The way is clear for drivers-in-chief all over the United States to aspire to error-free driving and the privilege of safe passage. But of the three human factors imperatives in the all-important DDT one alone – vehicle control – has been definitively improved. The other essential responsibilities – responding to the demands of the road system and the accommodation of other vehicles – are ignored by current technologies and are left to the unassisted driver. It is no longer sufficient to blame the driver and seek their replacement by machines. People have a continuing and enduring right to safe passage. The first step is the belated elevation of the national roadway infrastructure to the status of a cyber–physical system; followed closely by the ordination of the cardinals of human factors – their benevolent influence is sorely needed.

REFERENCES

1. National Highway Traffic Safety Administration (NHTSA) (2024) Overview of Motor Vehicle Traffic Crashes in 2022. Report DOT HS 813 560.
2. National Highway Traffic Safety Administration (NHTSA) (2022) Overview of Motor Vehicle Crashes in 2020. Report DOT HS 813 266.
3. Theeuwes, Jan (Ed.) (2017) *Designing Safe Road Systems: A Human Factors Perspective*. Routledge, CRC Press.
4. Mohaddes, Abbas & Sweatman, Peter (2016) Transformational Technologies in Transportation. Transportation Research Board. Circular E-C 208. May 2016.
5. Parry, Richard (2003) *Episteme* and *Techne. The Stanford Encyclopedia of Philosophy* (Spring 2024 Edition), Edward N. Zalta & Uri Nodelman (eds.).
6. Asimov, Isaac (1950) *Runaround. I, Robot (The Isaac Asimov Collection ed.)*. Doubleday.
7. Walker Smith, Bryant (2017) Automated Driving and Product Liability. Michigan State Law Review. 2017 MICH. ST. L. REV. 1.

Index

Printed in the United States
by Baker & Taylor Publisher Services